Results from the

Fourth Mathematics Assessment

of the

National Assessment of Educational Progress

edited by

Mary Montgomery Lindquist
Columbus College
Columbus, Georgia

National Council of Teachers of Mathematics

Copyright © 1989 by
THE NATIONAL COUNCIL OF TEACHERS OF MATHEMATICS, INC.
1906 Association Drive, Reston, Virginia 22091
All rights reserved

Second printing 1990

Library of Congress Cataloging-in-Publication Data

Results from the fourth mathematics assessment of the National
    Assessment of Educational Progress / edited by Mary Montgomery
    Lindquist.
        p.   cm.
    Bibliography: p.
    ISBN 0-87353-274-0
    1. Mathematics—Study and teaching—United States.
2. Mathematical ability—Testing. 3. National Assessment of
Educational Progress (Project) I. Lindquist, Mary Montgomery.
II. National Council of Teachers of Mathematics.
QA13.R48 1988
510'.71'073—dc20
                                                                                        89-3335
                                                                                          CIP

This material is based upon work supported by the National Science Foundation under Grant No. SPA-8652477. The Government has certain rights in this material. Any opinions, findings, and conclusions or recommendations expressed in this material are those of the author(s) and do not necessarily reflect the views of the National Science Foundation.

The publications of the National Council of Teachers of Mathematics present a variety of viewpoints. The views expressed or implied in this publication, unless otherwise noted, should not be interpreted as official positions of the Council.

Printed in the United States of America

CONTENTS

Preface   v

Foreword   vii
    John A. Dossey

1. Introduction   1
    Thomas P. Carpenter

2. Mathematical Methods   10
    Edward A. Silver and Thomas P. Carpenter

3. Discrete Mathematics   19
    Catherine A. Brown and Edward A. Silver

4. Data Organization and Interpretation   28
    Catherine A. Brown and Edward A. Silver

5. Measurement   35
    Mary M. Lindquist and Vicky L. Kouba

6. Geometry   44
    Mary M. Lindquist and Vicky L. Kouba

7. Variables and Relations   55
    Jane O. Swafford and Catherine A. Brown

8. Number and Operations   64
    Vicky L. Kouba, Thomas P. Carpenter, and Jane O. Swafford

9. Calculators   94
    Vicky L. Kouba and Jane O. Swafford

10. Attitudes   106
    Jane O. Swafford and Catherine A. Brown

11. What Can Students Do? (Level of Mathematics
    Proficiency for the Nation and Demographic Subgroups)   117
    *John A. Dossey, Ina V. S. Mullis, Mary M. Lindquist,
    and Donald L. Chambers*

12. Minority Differences in Mathematics   135
    Martin L. Johnson

13. Gender Differences in Mathematics   149
    Margaret R. Meyer

14. Summary and Conclusions   160
    Thomas P. Carpenter and Mary M. Lindquist

Bibliography   170

PREFACE

The National Assessment of Educational Progress completed its fourth mathematics assessment during the 1985–86 school year and finished the analyses of the results in 1988. This monograph, prepared by an interpretive team of the National Council of Teachers of Mathematics, represents a comprehensive discussion of the results of the fourth assessment based on the analyses provided by NAEP.

Both item statistics and scaling statistics are available for NAEP from the contractor, the Educational Testing Service. The scaling analyses provide information about overall performance and trends that make them especially useful in policy decisions. A complete discussion of these results can be found in *The Mathematics Report Card: Are We Measuring Up?* (Dossey, Mullis, Lindquist, and Chambers 1988).

This monograph presents examples of the cognitive and affective items and summarizes the results of the individual items for the different grade levels and subgroups. These discussions, along with the authors' interpretations, have been found by teachers, curriculum developers, and researchers to be most helpful. The monograph also discusses the scaling and offers an interpretation of the overall results of the assessment.

NCTM has worked cooperatively with NAEP throughout the four mathematics assessments by reviewing objectives and items and preparing reports. A complete listing of these reports appears in the Appendix. For this assessment, NCTM appointed a task force to plan and secure funds from the National Science Foundation (NSF) for the interpretive reports. Members of the task force were Mary M. Lindquist (chair), Thomas P. Carpenter, Vicky L. Kouba, Ann McAloon, and Gail Robinette. After a grant was received from NSF, the following interpretive team was appointed:

> Mary M. Lindquist, *Chair*
> Catherine A. Brown
> Thomas P. Carpenter
> Vicky L. Kouba
> Edward A. Silver
> Jane O. Swafford

Martin L. Johnson and Margaret R. Meyer also wrote chapters of the monograph, and John A. Dossey prepared the introductory remarks.

The following teachers reviewed the manuscripts for the monograph: Bridget Arvold, Jan Brown, Janet Barnard, Pamela Baer, Karl Byers, Dorothy Deeb, Marvin Doubet, Theresa Garrison, Carl Grady, Gina Griffin, Michael McGuigan, and Janice Roback.

Thanks are expressed to all who helped in the preparation of this document, to NAEP for their cooperation with the project, to NCTM for initiating the project and for publishing the various reports, and to NSF for the funding.

REFERENCE

Dossey, John A., Ina V.S. Mullis, Mary M. Lindquist, and Donald L. Chambers. *The Mathematics Report Card: Are We Measuring Up?* Princeton, N.J.: Educational Testing Service, 1988.

FOREWORD

The results of the mathematics assessments of the National Assessment of Educational Progress (NAEP) serve as one barometer of our nation's level of knowledge and skill in mathematics. The Fourth Mathematics Assessment, conducted in the 1985–86 school year, was no exception. Its major findings received a great deal of attention in the press and continued the current discussion of schooling in America. The media thus contributed to the growing awareness of Americans that the emphases in mathematics education needed to shift from the narrow view based on facts and computational procedures to a broader vision of mathematics that recognized the importance of problem solving.

Although these curricular changes have been the talk of mathematics educators at all levels since the publication of *An Agenda for Action* in 1980, the national attention paid to the results of the latest NAEP assessment comes at an especially fortunate time for mathematics teaching and learning. This renewed interest in mathematics continues the discussion that was ignited by the release of the results of the Second International Mathematics Study in 1987 and builds on the more generic discussion of educational reform in general. NCTM's *Curriculum and Evaluation Standards for School Mathematics* will continue this discussion and provide a basis for setting expectations for student performance that will help in addressing these past shortcomings.

Several of the findings of the present NAEP analysis suggest that the new emphases are having an effect. The 1990 NAEP Assessment and subsequent biennial assessments will serve as a gauge against which we can begin to measure our continued progress on righting the curriculum and redirecting our efforts toward more effective learning and performance on the part of American students.

Although the performance of students at all levels is considerably less than the curriculum would suggest, significant differences in the trend data indicate that 9-, 13-, and 17-year-olds are increasing their performance on the NAEP. Both the academic growth and increased enrollment patterns in secondary school mathematics of black and Hispanic youth suggest that some traditional barriers within mathematics education are being overcome and that the opportunity to study mathematics is becoming more common across the country.

The growing gap between the opportunity to learn and students' performance levels suggests that as a nation we must continue to support and assist students experiencing difficulty in learning mathematics. Mathematics specialists are as important to mathematics learning and achievement in the schools as reading specialists are to reading. Our society must recognize that mathematics is, along with reading, a critical factor in the breadth of students' opportunities and their final economic status in life. Not only must society recognize this fact and support students in their attainment of mathematics, but students themselves need to learn the value of mathematics in their future careers. Only 47 percent of the 17-year-olds in this NAEP sample thought that they would work in an area that would require some mathematics skills.

This volume offers an in-depth analysis of students' ability to function in the major areas of the mathematics curriculum. The NCTM/NAEP Interpretive Team that produced this book has carefully sifted the results and patterns of performance to give mathematics educators at all levels a picture of what students know and can do. This information provides a base for both research and in-service efforts aimed at making the teaching and learning of mathematics a more efficient process for teachers and students. The field of mathematics education is indebted to the writers on that team for offering their insights as we work to develop students and programs that will attain the new standards the profession is setting for mathematics education.

John A. Dossey
Past President, NCTM

1

INTRODUCTION

Thomas P. Carpenter

THE National Assessment of Educational Progress (NAEP) was established in 1969 to provide information on the educational performance of American youth and to measure changes in performance over time. Since its inception NAEP has gathered information about the performance of 9-year-old, 13-year-old, and 17-year-old students in mathematics, writing, reading, science, social studies, and other disciplines. Assessments in mathematics were conducted during the school years ending in 1973, 1978, 1982, and 1986. This report examines the results for the fourth mathematics assessment conducted in 1986.

OBJECTIVES AND ITEM DEVELOPMENT

The NAEP attempts to be responsive to changes in the mathematics curriculum. For every assessment the objectives that guide the development of items are reviewed and revised as appropriate. The objectives that provided a framework for the fourth mathematics assessment were written by a panel of five mathematics educators: James Bruni, Iris M. Carl, Clyde L. Corcoran, F. Joe Crosswhite, and Shirley Hill. Their recommendations were based on an initial review of the objectives of the previous assessment by twenty-five mathematics educators and classroom teachers. A draft set of objectives prepared by the panel was reviewed by another group of twenty-five mathematics educators and classroom teachers and revised on the basis of recommendations from this survey.

The mathematics objectives cover seven broad content areas:

1. Fundamental methods of mathematics
2. Discrete mathematics
3. Data organization and interpretation
4. Measurement
5. Geometry
6. Relations, functions, and algebraic expressions
7. Numbers and operations

Each of the content areas was assessed at five levels:

A. Problem solving and reason
B. Routine application
C. Understanding and comprehension
D. Skill
E. Knowledge

The relationship between content areas and process levels is represented in figure 1.1.

	CONTENT						
PROCESSES	(1) Fundamental methods of mathematics	(2) Discrete mathematics	(3) Data organization and interpretation	(4) Measurement	(5) Geometry	(6) Relations, functions, algebraic expressions	(7) Number and operations
(A) Problem solving/ reasoning							
(B) Routine application							
(C) Understanding comprehension							
(D) Skill							
(E) Knowledge							

Fig. 1.1. Objectives framework for the Fourth Assessment

Content Areas

The content included in the fourth mathematics assessment was based primarily on the current curriculum of elementary and secondary schools. Because the assessment is not designed to administer different items to students with different mathematics backgrounds, the objectives focused on content that had been covered by a majority of students at a given grade level. As a consequence few items were included that would have required a formal course in geometry, advanced algebra, or precalculus. Each of the content categories is described briefly below. More complete descriptions are provided in the chapters in which the results for each topic are discussed. The distribution of items by content area is summarized in table 1.1.

INTRODUCTION

Table 1.1
Number of Items by Content Category

Category	Grade 3	Grade 7	Grade 11
Fundamental methods of mathematics	17	23	36
Discrete mathematics	3	12	17
Data organization and interpretation	16	23	23
Measurement	27	49	45
Geometry	6	26	41
Relations, functions, and algebraic expressions	8	14	47
Number and operations	56	146	132
Calculators*	11	31	32

*Items included in other content categories for which calculators were available.

Fundamental methods of mathematics. This category, new to this assessment, focused on basic mathematical processes that cut across content areas. It included items that assessed the use of basic problem-solving strategies, the use of models, the mathematical representation of problem situations, a general understanding of the nature of proof and axiomatic systems, and logic.

Discrete mathematics. Discrete mathematics is a new content category that did not appear in the list of content areas assessed in previous assessments; however, most of the topics in this category were included in earlier assessments. This category included such topics as probability, combinations, and permutations.

Data organization and interpretation. This category was also identified as a separate content area for the first time in the fourth assessment. As with discrete mathematics, essentially the same topics were covered in previous assessments, but they were lumped together with a variety of other items in a category called "Other Topics." The data organization and interpretation items assessed the ability to read and interpret graphs, charts, and tables and the knowledge of measures of central tendency and distribution.

Measurement. The items in this content category included selecting appropriate units; identifying the equivalence of units; using rulers and other measuring instruments; and finding perimeter, area, and volume. Both metric and customary units of measure were included.

Geometry. With a few exceptions the geometry items could be answered on the basis of the informal geometry taught in elementary school and junior high school mathematics classes. Items in this category assessed the recognition of common geometric figures, the properties of figures, congruence, symmetry, the Pythagorean relation, spatial visualization, and the like.

Relations, functions, and algebraic expressions. The items in this category included basic principles of variables that are taught in elementary and middle school as well as topics that are specific to a formal course in algebra. The items dealt with such topics as using variables to represent situations and relationships, simple functions, solving equations and inequalities, and simplifying algebraic expressions.

Number and operations. The items in this category included numeration, basic number concepts, computation, and the application of computational skills to solve problems.

Process Levels

The five process levels were identified to encourage the inclusion of a broad range of items that assessed different levels of thinking. Although items were assigned to different process levels, the assignment of items to process level was much more subjective than the assignment of items to content categories. The process levels generally guided the development of items, but little attempt was made to validate systematically the resulting process categories.

Problem solving. This category was intended to include items that assessed higher-order thinking that involved the integration of concepts and skills to solve problems for which there was no clear method of solution. Items at this level could not be solved by the routine application of a concept or skill.

Routine application. Items at this level involved the use of mathematical knowledge or skills in familiar situations with which it is presumed that students have had experience. Most items were typical textbook problems or problems that students might be expected to encounter frequently in their daily lives. Although the mathematical operations were not specified, the identification of the appropriate procedure was almost automatic.

Understanding and comprehension. Items assessing understanding focused on basic mathematical concepts and principles. They frequently required students to identify or establish relationships between different representations, like identifying which model represented a given number sentence.

Skill. Mathematical skill refers to routine mathematical manipulations that have been learned and practiced. Items that measured skills did not require students to decide what operation to use, and the objective for these items was to assess proficiency in using an algorithm rather than the understanding of how it works. Items that assessed mathematical skills required the performance of such procedures as making measurements, multiplying fractions, solving an equation, or reading a table.

Knowledge. Mathematical knowledge refers to the recall or recognition of mathematical ideas, figures, or symbols. Knowledge items included such tasks as identifying common geometric figures or recalling a number fact.

INTRODUCTION 5

Other Areas in the Assessment

In addition to the topics represented by the content-by-process matrix described above, several additional areas were covered in the fourth assessment.

Since the second assessment, NAEP has assessed students' attitudes toward mathematics. Four categories of attitude measures were included in the fourth assessment: (1) mathematics in school, (2) mathematics and oneself, (3) mathematics and society, and (4) mathematics as a discipline.

Students were asked about the availability and use of calculators and computers. The assessment also included a subset of computation and routine application items for which a calculator was used.

METHODOLOGY

The first three mathematics assessments were conducted by the Education Commission of the States. The fourth mathematics assessment was conducted by the Educational Testing Service. The change in administration has been accompanied by some changes in methodology that have significant consequences for the interpretation of the results. These changes are discussed in the sections and chapters that follow. For more detail on the methodology, the reader should consult *Implementing the New Design: The NAEP 1983–1984 Technical Report* (Beaton 1987).

Exercise Development

Using the framework of content categories and process levels outlined above, a group of mathematics educators worked with the staff of the Educational Testing Service to develop the items for the assessment. The items passed through a series of stages of development, review, and field testing. After items were written, with input from mathematics educators, they were reviewed internally by ETS for such things as content and sensitivity. After initial field testing, they were reviewed by the United States Department of Education and others. The content specialist team then selected the final set of items, which was again reviewed using the same procedure as the initial set.

A number of items from previous assessments were not released to the public. Many of these items were included in the fourth assessment to provide continuity and to establish a basis for measuring change in performance from previous assessments. (See chapter 11 for a discussion of the measurement of change.) The number of newly developed items and the number of items from previous assessments at each grade level are reported in table 1.2.

There are some differences between the new items and the items from previous assessments. Items from previous assessments frequently were

Table 1.2
Number of Cognitive Items in NAEP Assessments

Assessment	Grade 3	Grade 7	Grade 11
Third and fourth	85	183	191
Fourth only	59	129	182

open ended. When multiple-choice items were used, the number of responses varied from as few as two to as many as seven, and most multiple-choice items included an "I don't know" response category. The newly developed items were all multiple choice and always included exactly four choices. None of the new items included "I don't know" as a possible response.

Sample

Matrix sampling procedures were used to identify a representative national sample of 3rd-grade, 7th-grade, and 11th-grade students. All together 18 033 3rd-grade students, 23 527 7th-grade students, and 31 938 11th-grade students participated in the assessment. In previous assessments, subjects were selected by age rather than grade, so the results for this assessment provide a somewhat different perspective on performance than the first three assessments. To give broad coverage of topics, item sampling procedures were used so that each student received approximately 10–15 percent of the items administered at each grade level. Each item was administered to approximately 2000 students at each grade level at which it was given.

Administration Procedures

The items were divided into blocks for administration at each grade level. There were seven blocks at the 3rd grade, nine at the 7th grade, and eleven at the 11th grade. At the 3rd grade there were about 20 cognitive and 5 attitude items in each block, and at the 7th and 11th grades there were approximately 35 cognitive and 15 attitude items in each block. A total of 144 cognitive and 39 attitude items were included in the seven 3rd-grade blocks. The 7th-grade and 11th-grade blocks included 312 and 373 cognitive items, respectively. There were 119 attitude items in the 7th-grade blocks and 157 attitude items in the 11th-grade blocks.

Each student who participated in the assessment was administered three blocks. The mathematics assessment was conducted concurrently with the science, reading, and computer assessments. The subjects were integrated, so that a student may have taken no mathematics blocks or as many as three mathematics blocks. Each block appeared in at least five different combi-

INTRODUCTION 7

nations in different test booklets. All together there were 46 booklets at the 3rd grade, 62 at the 7th grade, and 86 at the 11th grade.

Students were allowed 16 minutes for each block, and students worked at their own pace through each block. This procedure is different from previous assessments. In the first three assessments, the administration of each item was accompanied by a paced audio recording in which the text of the item was read to the student. Students did not work on a given item until they were told to do so, and the time for each item was controlled. The administration procedures for the first three assessments were designed to provide as accurate measures as possible on individual items. Because of the speeded nature of the fourth assessment, performance on individual items is a little more difficult to interpret. For each block, some students did not complete all the items, and in some cases the completion rate was as low as 25 percent. Because the items appeared in a specific order within each block, a number of items had relatively high nonresponse rates. The consequences of these missing data are discussed below.

The fourth assessment was administered by trained test administrators to groups of approximately thirty-five students. In previous assessments, all the students in each administration group received the same test items. To increase power, different students within each administration group in the fourth assessment received different test booklets. As a consequence, items were not read to students, and each student worked at his or her own pace.

In previous assessments, students in different grade levels were tested at different times during the year. In the fourth mathematics assessment, all students were tested from 17 February to 2 May. Procedures that were used to administer and equate items from the previous assessments will be discussed in chapter 11.

Reporting and Interpreting Results

The concept of a total score on the mathematics assessment is inappropriate. Such a score would have little meaning and would be seriously influenced by the specific items included in the assessment. Previous assessments have been reported on an item-by-item basis. This level of reporting has been particularly useful to researchers, curriculum developers, and teachers who were seeking specific information about performance in particular areas. Because of the value of item-by-item analysis for making decisions about curriculum, this report continues the practice of discussing performance on individual items, in spite of the limitations of the data created by the new administration procedures. The current data appear to provide reasonable estimates of performance on individual items if one does not assume they are accurate to the nearest percent. In general we are concerned with relative performance on different types of items and the identification of patterns of performance and critical areas of concern. The item data appear to provide reasonable estimates for these purposes. If performance on a given item is at the 30 percent level, one can draw reasonable conclusions about the success of the curriculum in teaching students to solve the given item without worrying too much about measurement error.

Only a subset of the items administered in the assessment is discussed in this report. Items were selected to illustrate performance, and the items not included support the conclusions reported. The items that are reported have been altered slightly so that they can be administered in future assessments to measure change. Care has been taken to ensure that the altered items require all the same skills and concepts as the original items.

Although item-by-item analysis yields a great deal of useful information, the amount of detail makes it difficult to draw broad general conclusions about overall performance and how it has changed over time. In the past, NAEP attempted to provide some summary of the data by aggregating them over content areas or process levels at each grade level. This procedure did not prove to be entirely satisfactory, since the aggregated scores had little meaning. For the current assessment, NAEP has attempted to construct scales that provide a profile of performance within different content areas. These scales are discussed in the chapter on scaling (chap. 11).

The Problem of Missing Data

The fact that students did not respond to some items presented to them creates a problem for the analysis of performance on an item-by-item level. Does one assume that students who did not respond to a given item would perform at the same level as the students who responded to that item, or does one assume that the students who did not respond to a given item did not respond because they could not do the item correctly? Neither assumption is completely warranted. It would be inappropriate to conclude that students who did not get to an item could not solve it, but it would also be inappropriate to conclude that students who work more slowly than other students would be as successful in solving a given item as the students who worked faster.

The nonresponses fall into two categories: items in the middle of the test that students did not answer and items at the end of the test that students did not reach. Most of the nonresponses fall into the latter category. Only a small number of items with nonresponse rates exceeding 5 percent preceded items to which students did respond. However, a sizable number of items near the ends of blocks had nonresponse rates exceeding 50 percent.

For this report, performance on an individual item is reported for students who responded to the item. This means that the statement "56 percent of the students correctly answered a given item" should be interpreted to mean that 56 percent of the students who responded to the item answered it correctly. It does *not* mean that if all students in the given grade had been given time to respond to the item, 56 percent would have responded correctly.

If the item had a high response rate, the difference between the two interpretations is academic, but if the response rate was relatively low, the reported level of performance may provide a less accurate estimate of performance for the population as a whole. The available data suggest that the reported levels of performance probably overestimate the performance of the population as a whole. Limited empirical evidence suggests that the performance of students who did not get to an item would be about .8 of the performance of the students who responded to the item. The net effect of applying this formula to the given data would be to reduce the estimate of performance by .02 to .10 of the performance level of students who responded to the item for nonresponse rates ranging from .10 to .50. This does *not* mean that the estimates would be reduced by 2 to 10 percentage points. For example, a performance level of 40 percent with a response rate of .90 would be reduced to 39 percent and a performance level of 40 percent with a response rate of .50 would be estimated at 36 percent for the population as a whole.

Adjusting performance for nonresponse rate would be based on limited empirical evidence and questionable assumptions. We have chosen to report the performance of the students who actually responded to the item and provide response rates for items in the tables. In most cases we have not discussed items with response rates lower than .50. In some cases this has limited our item selection. Many of the unusual nonroutine items or items that required specific mathematical coursework tended to be toward the end of blocks, and the response rates for these items were frequently relatively low.

REFERENCE

Beaton, Albert E. *Implementing the New Design: The NAEP 1983–1984 Technical Report.* Princeton, N.J.: Educational Testing Service, 1987.

2

MATHEMATICAL METHODS

Edward A. Silver   *Thomas P. Carpenter*

IN THE 1985–86 mathematics assessment, NAEP incorporated a new content category, mathematical methods. According to the test designers (NAEP 1986), the items in this category deal with "those processes that are central to the extension and development of mathematics and to its use." The designers intended the category to include concepts and processes of deductive and inductive proof, logic, mathematical models, structure and system, routine procedures, problem-solving strategies, and empirical induction. However, the coverage on the 1985–86 assessment in some of these areas was rather thin. For example, only one item measured empirical induction, and that item was flawed. Nevertheless, the fifty-two items in this category, together with a few related items found elsewhere in the assessment, provide some information about the knowledge that students at the three grade levels appear to have about some fundamental mathematical processes and concepts.

HIGHLIGHTS

• Students at all grade levels tended to reason better on items that involved familiar, everyday settings than on items that involved abstract contexts.

• Students at all three grade levels performed poorly on items that involved the coordination of more than two simultaneous problem constraints.

• When they were specifically asked, older students generally could recognize problems that contained insufficient information and could identify missing information required to solve a problem. However, many students also tended to indicate insufficiency of information when sufficient information was given.

MATHEMATICAL METHODS

- Most 11th-grade students demonstrated little understanding of the nature and methods of mathematical argumentation and proof.
- Students in grade 11 who had completed at least one year of college-preparatory mathematics performed significantly better than those students who had taken no college-preparatory mathematics course. But, 11th-grade students who had completed three years of college-preparatory mathematics generally performed only slightly better than all students who had completed one year.

DISCUSSION OF RESULTS

In this chapter, the discussion of results is organized around four general themes: logical reasoning, routine procedures, problem solving, and proof. These general themes encompass almost all the items dealing with mathematical methods.

Logical Reasoning

Several items were included that assessed students' logical-reasoning abilities. Table 2.1 contains results for two items that assessed the reasoning abilities of students at all three grade levels. It is interesting to note that unlike performance on almost every other item administered at all three grade levels, the performance on the first item did not improve with age. Older students did only slightly better, if at all, than younger students on

Table 2.1
Reasoning in a Nonmathematical Context

	Percent Responding[a]		
Item	Grade 3	Grade 7	Grade 11
Everyone on the team is tall.			
A. If Tom is short, then—			
Tom is on the team.	5	1	1
Tom is not on the team.*	62	62	68
There is not enough information to tell if Tom is on the team.	35	35	30
I don't know.	2	2	1
B. If Jane is tall, then—			
Jane is on the team.	68	58	47
Jane is not on the team.	5	3	2
There is not enough information to tell if Jane is on the team.*	13	38	50
I don't know.	14	1	1

[a] The response rates for item A were as follows: grade 3, .78; grade 7, .97; and grade 11, .95. The response rates for item B were as follows: grade 3, .76; grade 7, .96; and grade 11, .95.
*Indicates correct answer.

this reasoning item. Moreover, almost a third of the students at all grade levels indicated that there was insufficient information to draw a conclusion. Apparently, there is little in students' school experience after 3rd grade, either in mathematics or in other subject areas, that positively affects their ability to reason in this kind of hypothetical situation. This observation is further supported by data indicating an increase of only 4 percent in the performance of students who had taken three years of college-preparatory mathematics over the performance of those who had taken only one year.

On the second item, which essentially involved recognizing the nonequivalence of a hypothetical statement and its converse, performance did improve substantially with age. Nevertheless, only about half of the 11th-grade students were able to answer this item correctly. Moreover, only 55 percent of the 11th-grade students who had completed or were enrolled in second-year algebra were able to answer this question successfully. Given the high percentage of students incorrectly choosing the same answer option on the first item, one suspects that many may have chosen that answer option, which was the correct answer for the second item, for the wrong reason.

Table 2.2 presents the results of performance on an item that essentially involved the recognition and application of the transitivity property of the "less than" relation in an abstract mathematical setting. It is interesting to

Table 2.2
Reasoning in a Mathematical Context (Grade 11)

Item	Percent Responding[a]

The letters in the diagram above represent numbers. If $x \to y$ means $x > y$, which of the following is NOT necessarily true?

$b > d$	16
$e > d$*	55
$a > d$	20
$a > c$	10

[a] The response rate was .71.
*Indicates correct answer.

contrast performance on this item, on which about one-fourth of the 11th-grade students chose answer options that reflected nontransitivity, with that on another item, on which nearly nine out of ten 11th-grade students appeared to recognize a straightforward application of the transitivity of the "older than" relation in a nonmathematical context. In general, student performance was higher when the task called for reasoning in a real-world setting rather than similar reasoning in a mathematical or otherwise abstract context. Knowledge that was available in nonmathematical settings was not readily transferred to mathematical contexts to solve related problems.

Routine Procedures

Some new items were added to this assessment to measure students' understanding of a few routine mathematical procedures. The general finding is that although students may be able to use routine procedures in school mathematics, they appear not to have gained an understanding of those procedures.

One item assessed students' understanding of both the process for finding common denominators and the conditions under which it is appropriate to do so when manipulating common fractions. Only about four of every ten 11th-grade students were successful on that item. The results for another of the items assessing understanding of routine procedures are presented in table 2.3. This item dealt with students' understanding of the procedure for multiplying by a fraction. Multiplication of a whole number by a fraction is not a serious computational difficulty for most 7th-grade and 11th-grade students. For example, about three-fourths of the 11th-grade students and about 60 percent of the 7th-grade students were successful on a related item that required only a straightforward computation of the product of a fraction and a whole number. Therefore, it is interesting that performance on the item shown in table 2.3 was so poor. The poor

Table 2.3
Understanding Multiplication by a Fraction

Item	Percent Responding[a]	
	Grade 7	Grade 11
Which is one way to find $\frac{3}{4}$ of a number?		
Divide by 3 and multiply by 4.	19	11
Divide by 4 and multiply by 3.*	26	38
Divide by 3 and divide by 4.	25	13
Multiply by $\frac{4}{3}$.	31	39

[a] The response rate was .70 for grade 7 and .96 for grade 11.
*Indicates correct answer.

performance may reflect a lack of conceptual understanding of fractions as operators, or it may result from a general confusion regarding multiplication and division of fractions. Response frequencies for each of the four answer choices were suspiciously close to the level of chance for this item. It is quite remarkable, and somewhat depressing, that even the best-prepared mathematics students in the sample had difficulty with this item. Fewer than half of the 11th-grade students who had completed second-year algebra were able to answer correctly this question about a simple mathematical procedure.

Problem Solving

In this assessment, there were a few items that measured some process aspects of mathematical problem solving. For example, one crucial aspect of problem solving involves the related processes of (a) determining whether or not there is sufficient information to solve a problem and (b) identifying needed information when more is needed. One item, which requested students to specify the missing information that would be needed to solve a problem, was answered correctly by about three-fourths of the 7th-grade students and fewer than half of the 3rd-grade students. On another item, which contained sufficient information, over half of the 3rd-grade students incorrectly indicated that there was insufficient information given to solve the problem. Referring to the first item in table 2.1, we see that about one-third of the students at all grade levels indicated that there was insufficient information to answer that item. In general, the data from all the items of this type suggest that when they were specifically asked to do so, many of the 3rd-grade students and most of the 7th-grade and 11th-grade students were capable of both recognizing problems that contained insufficient information and identifying needed information when it was required to solve a problem. Whether or not this success signifies insightful problem-solving behavior is unknown, however, since many students also tended to indicate insufficiency of information when a problem could be solved using the given information.

Another important problem-solving process assessed by a few items on the test was the ability to associate a problem with an appropriate mathematical model of the problem situation, such as a diagram or a graph, or to recognize the equivalence to two different models for the same problem. About half of the 3rd-grade students were able to match a numerical expression with a picture, and about one-third were able to find a number-line representation that was equivalent to a given number sentence.

Table 2.4 presents results from two items that illustrate the difficulty that students at all grade levels appeared to have in coordinating complicated problem information, especially when it was presented as simultaneous problem constraints. On the first item, which is very elementary, four of

MATHEMATICAL METHODS 15

Table 2.4
Coordinating Simultaneous Problem Constraints

Item	Percent Responding[a]		
	Grade 3	Grade 7	Grade 11
A. What number am I?			
(i) If you add digits, you get 15.			
(ii) My tens digit is one more than my hundreds digit.			
(iii) My ones digit is one more than my tens digit.			
834	NA[b]	NA	11
780	NA	NA	16
678	NA	NA	12
456*	NA	NA	60
B. Four cars wait in a single line at a traffic light. The red car is first in line. The blue car is next to the red. The green car is between the white car and the blue car. Which color car is at the end of the line?			
Red	6	3	NA
Blue	56	49	NA
Green	10	4	NA
White*	29	45	NA

[a] The response rate for item A was .96 for grade 11. The response rate for item B was .88 for grade 3 and .98 for grade 7.
[b] NA indicates that the item was not answered by students at that grade level.
*Indicates correct response.

every ten 11th-grade students chose an answer that had only two of the three specified characteristics, thereby ignoring one of the problem constraints. It would appear from the responses to several items on the assessment that 11th-grade students were generally capable of coordinating two simultaneous problem constraints, but their performance deteriorated markedly when three or more constraints were presented. On the second item in table 2.4, it appears that many 3rd-grade and 7th-grade students considered only the first and third problem constraints, thereby leading them to choose "blue" as the color of the last car in line, despite the fact that the second constraint clearly prevented that ordering.

When items actually required students to solve fairly complex problems, performance was predictably low. For example, on an open-ended multiple-step problem requiring multiplication and subtraction of two-digit whole numbers, fewer than a third of the 11th-grade students and fewer than a tenth of the 7th-grade students answered correctly. Students were more successful in answering problem-solving items when they were presented in multiple-choice rather than open-ended form. Moreover, they were more successful when the item was worded in such a way as to allow working back and forth between the item stem and the answer choices. Table 2.5 presents the performance of 7th-grade and 11th-grade students on two items that illustrate this phenomenon. Although performance was not very strong on either item at either grade level, it was better on the first item, on which it is possible to move back and forth between the problem constraints and the answer choices, evaluating each choice until a correct answer is found.

Table 2.5
Solving Moderately Complex Problems

	Percent Responding[a]	
Item	Grade 7	Grade 11
A. When the items in a box are put in groups of 3 or 5 or 6, there is always 1 item left over. How many items are in the box if there are fewer than 50?		
16	14	9
29	28	19
30	22	9
31*	36	63
B. Suppose you have 8 coins and you have at least one of each of a quarter, a dime, and a penny. What is the *least* amount of money you could have?		
36¢	37	41
41¢*	23	43
43¢	25	7
72¢	15	9

[a] The response rate for item A was .54 for grade 7 and .95 for grade 11. The response rate for item B was .48 for grade 7 and .88 for grade 11.
*Indicates correct response.

In the second problem, however, it is almost impossible to adopt that strategy; it is much more reasonable and efficient to work forward from the given problem information to find a solution and then look to find it in the set of choices given. On this item, and on similar items that did not permit working back and forth, student performance was generally lower than when it was possible to work backward from the answer choices.

Proof and Proof-related Methods

The assessment also included several items that dealt with students' understanding of the standard methods of formal argumentation and proof in mathematics. Since the ability to construct arguments that prove statements within a mathematical system, rather than simply being able to produce examples and demonstrations, is fundamental to the discipline of mathematics, students' performance on these items provides some indication of the extent to which the fundamental nature and methods of proof are being communicated in the school mathematics curriculum and current school practice.

One important aspect of mathematical argumentation is the production of examples and counterexamples. Table 2.6 presents the performance of 11th-grade students on two items that required the recognition of a counterexample. Neither the algebra questions nor the geometry question was correctly answered by a majority of the 11th-grade students. The poor performance on these items suggests that most students do not learn to recognize counterexamples from their school mathematics experience.

Two items assessed 11th-grade students' understanding of the terms *axiom* and *theorem* as well as the basic nature of a mathematical proof. The majority of the students responded that a theorem was essentially a demon-

Table 2.6
Recognizing a Counterexample (Grade 11)

Item	Percent Responding[a]
A. Larry says that $n^2 \geq n$ for all real numbers. Of the following, which value of n shows the statement to be FALSE?	
$-1/2$	23
0	29
1/10*	39
1	9
B. Jim says, "If a 4-sided figure has all equal sides, it is a square." Which figure might be used to prove that Jim is wrong?	
□ (square)	15
◢◣ (trapezoid)	31
⬠ (pentagon)	24
▱ (rhombus)	31

[a] The response rate was .91 for item A and .54 for item B.
*Indicates correct response.

stration or an assumption, and fewer than one-fourth of the students were able to identify correctly the meaning of an axiom. Beyond the difficulty that students might have with the technical vocabulary involved, the poor performance on these items suggests that many 11th-grade students are confused about the fundamental distinctions among mathematical demonstrations, assumptions, and proofs.

Another item measured students' understanding of the important methods of indirect proof, or proof by contradiction. Despite the relatively straightforward approach taken in this question, only about one-third of the 11th-grade students correctly answered the questions. The responses of 11th-grade students on another item, which assessed their understanding of the method of mathematical induction, showed similarly poor understanding of the basic procedure.

Since many students never come in contact with a proof in their elementary and high school mathematics courses, these results might be expected. It is true that 11th-grade students who had completed a college-preparatory geometry course performed significantly better on the items dealing with mathematical proof than students who had not taken geometry. Nevertheless, performance on these items was generally only slightly better for students who had taken geometry than it was for students who had simply completed Algebra 1. Since secondary school geometry is usually the course in which mathematical proof and proof-related methods are demonstrated and discussed, one might have expected a more substantial difference in performance. Moreover, the general level of success on these items was quite low: except for one, the items were correctly answered by fewer than half of the 11th-grade students who had taken two or three years of college-preparatory mathematics.

The generally poor performance on these items dealing with proof and proof-related methods suggests the extent to which students' experiences in school mathematics, even for students in college-preparatory courses, may often fail to acquaint them with the fundamental nature and methods of the discipline.

REFERENCE

National Assessment of Educational Progress. *Math Objectives: 1985–86 Assessment.* Princeton, N.J.: Educational Testing Service, 1986.

3

DISCRETE MATHEMATICS

Catherine A. Brown   *Edward A. Silver*

THE school mathematics curriculum shows a heightened awareness of the widening role of probability, permutations, and combinations in modeling situations that occur in many disciplines. A number of items on the assessment dealt with these topics from discrete mathematics. Unfortunately, student knowledge of many other topics that are generally considered to be discrete mathematics was not assessed. For example, no items related to graph theory and only one item dealing with linear algebra appeared on the assessment.

HIGHLIGHTS

• Most 3rd-grade students had difficulty with simple probability items.

• Students in the 7th and 11th grades generally had difficulty with probability items, except those items involving very basic, simple probability.

• Performance on probability items suggests that some students relied on faulty intuitive notions of probability rather than formal knowledge of the subject matter.

• Most 7th- and 11th-grade students were unable to do items involving permutations or combinations.

DISCUSSION OF THE RESULTS CONCERNING DISCRETE MATHEMATICS

Many current lists of basic skills in mathematics include topics in discrete mathematics. Probability, permutations, and combinations are appearing in the newer textbook series as early as elementary school. However, many teachers and curriculum developers consider them luxury topics and either omit them or leave them until late in the school year, to be covered if time permits. Similarly, topics from linear algebra are making their way into the high school textbooks only to be omitted or lightly covered by the teacher. Although it is widely recognized that these topics in the school mathematics curriculum are among the most useful in applications outside the classroom, they are not often taught.

The fourth national assessment included a number of exercises that addressed probability, permutations, and combinations. Thus, we can document current performance in these areas at all three grade levels assessed. Unfortunately, many of the items on the assessment that addressed these topics appeared near the end of the item blocks. The conclusions drawn here are based on data for items that had response rates of 50 percent or better. Response rates for items given in the tables are reported in the tables.

Probability of Events

Simple events. Students at all three grade levels seem to have some understanding the probability of simple events, although responses to assessment items indicate that older students had a better understanding than younger students. For example, the item in table 3.1 was presented to

Table 3.1
Situation Providing Greatest Chance

Item	Percent Responding[a]	
	Grade 3	Grade 7
There is only one blue chip in each of the bags shown below. Without looking you are to pick a chip out of one of the bags. Which bag would give you the greatest chance of picking the blue chip?		
Bag with 10 chips*	38	72
Bag with 100 chips	3	2
Bag with 1000 chips	35	19
It makes no difference.	10	6
I don't know.	13	1

[a] The response rate was .89 for grade 3 and .99 for grade 7.
* Indicates correct response.

both 3rd- and 7th-grade students. Students in grade 7 appear to have some knowledge of the effect of the size of a sample space on the probability of an event. The rate of correct responses by 3rd-grade students, however, indicates a lack of understanding of this effect. The fact that a large percentage of 3rd-grade students selected the foil representing the bag that would give one the least chance of picking the blue chip suggests that students may realize that the size of the sample space affects the probability of an event but are unclear about the direction of that effect. Many 3rd-grade students selected the "I don't know" foil, which may reflect minimal formal instruction on these topics before this grade level.

The results for two other items, one presented to 3rd-grade students and a similar one presented to 7th- and 11th-grade students, offer insight into children's understanding of the probability of simple events. When 3rd-grade students were asked to determine the chances of picking one of three different objects, about a third of the students responded that they did not know the answer and about a third answered correctly. The item presented to 7th-grade and 11th-grade students was somewhat more difficult but still relatively uncomplicated: In a multiple-choice format, these students were asked to select the probability of picking a red object from a jar containing 2 red and 3 blue objects. About half of the 7th-grade students and two-thirds of the 11th-grade students answered correctly. The foil 2/3 was selected by about a fourth of the 7th- and 11th-grade students, perhaps because those were the numbers given in the problem. (It should be noted that 67 percent of the 7th-grade and 23 percent of the 11th-grade students did not respond to the items.)

Further evidence of the weakness of students' grasp of probability notions is found in the responses to an item that asked students at both upper-grade levels to chose the probability that a girl with three pennies in her pocket and nothing else will get a penny if she takes a coin from her pocket. About one-fourth of the students selected 1/3, again selecting the foil that involved the numbers given in the problem. Only about half of the 7th-grade students and two-thirds of the 11th-grade students selected the correct response of 1.

Calculation of the probability of an event not occurring seems to be more difficult for students than calculating the probability of the event occurring. The items in table 3.2 were administered to both 7th- and 11th-grade students. Most students knew how to find the probability of the very simple event in item A, but when asked to find the probability of the complement of that event (item B), they were far less successful. Note that both the "I don't know" foil and the foil that listed the probability from the previous item were frequently selected by students. These results indicate that students may not understand the relationship between the probability of an event and the probability of the complement of this event. They also suggest

Table 3.2
Calculation of Probability of Simple Events

Item	Percent Responding[a]	
	Grade 7	Grade 11
A. Patty is rolling a number cube with 1, 2, 3, 4, 5, and 6 dots on its faces. What is the probability of Patty getting a 2 on her next roll?		
0	7	1
1/6*	38	76
2/6	5	3
3/6	6	3
4/6	4	6
5/6	27	2
I don't know.	14	8
B. Patty rolls the number cube again. What is the probability of Patty *not* getting a 2 on this roll?		
0	7	2
1/6	38	20
2/6	5	5
3/6	6	5
4/6	4	6
5/6*	27	55
I don't know.	14	9

[a] The response rate for item A was .98 for grade 7 and .99 for grade 11. The response rate for item B was .98 for grade 7 and .99 for grade 11.
* Indicates correct response.

that students may have difficulty with items involving compound events, since this is related to another method by which they could have answered item B.

Compound events. Items involving compound events were given only to 11th-grade students. Their intuitive procedures for dealing with probability seem to create difficulty with compound events. For example, one item asked students to determine the chances that at least one tail will appear if two fair coins are tossed. Only 5 percent of the students answering the item selected the correct answer of "3 in 4," whereas 70 percent of the students responding selected "1 in 2," a foil that used the two numbers given in the exercise and that they perhaps intuitively associated with experiments involving coin tossing.

The results on items for the 11th-grade students support the conclusion that compound events cause difficulty for students. The distribution of responses for the two items in table 3.3 indicates that many students may have been guessing. However, intuitive or relatively weak understanding of probability might also explain some of the responses. In item A, for example, 1/12 may have been selected by a number of students as the

Table 3.3
Calculation of Probability of Compound Events by Grade 11 Students

Item	Percent Responding[a]
A. A jar contains 4 green, 6 blue, and 8 yellow cubes. One cube is drawn from the jar. What is the probability that the cube drawn at random is yellow or green?	
1/12	24
1/8	16
1/2	16
2/3*	44
B. A coin is tossed and a die is rolled. What is the probability that the coin comes up tails and the die comes up 5?	
1/12*	34
1/8	25
1/5	18
6/12	22

[a]Response rate was .88 for item A and .84 for item B.
*Indicates correct response.

correct response because they incorrectly reasoned that since there are 4 + 8 favorable outcomes in the experiment, the probability should be found using 1 and the number of favorable outcomes. Students may believe that the probability of an event depends on the number of favorable outcomes and the number of possible outcomes but have difficulty expressing this in the conventional format. Similarly, about a fourth of the students may have selected 1/8 as the correct response in item A in the belief that there were 6 plus 2 possible outcomes and thus 8 should be used as the denominator of the unit fraction.

It seems that students are also unaware of circumstances that might influence the probability of compound events. For example, responses to the item shown in table 3.4 indicate that many 7th-grade and 11th-grade students do not understand how or when the probability of such an event may be influenced by the order of selection. About one-third of the 7th-grade students and one-half of the 11th-grade students answered this exercise correctly. However, the number of answers that were consistent with the belief that the order of selection influences probability in this situation indicates that misunderstanding may be quite common. These responses may also reflect the belief that the event of selecting from one bowl is not independent from the event of selecting from the other bowl.

Independent events. The failure to recognize the independence of certain events is common at all the grade levels assessed. An item related to the two

Table 3.4
Effect of Order of Selection on Probability

Item	Percent Responding[a]	
	Grade 7	Grade 11
Cards in glass bowl: $\boxed{8}\ \boxed{8}\ \boxed{2}\ \boxed{8}\ \boxed{2}\ \boxed{6}$		
Cards in plastic bowl: $\boxed{2}\ \boxed{3}\ \boxed{3}$		
As shown above, some numbered cards are placed in two bowls. They are placed so that the numbers cannot be seen. If a student were to select one card at random from each bowl, which of the following would affect the probability of selecting both a $\boxed{2}$ and a $\boxed{3}$?		
Select from the glass bowl first.	15	10
Select from the plastic bowl first.	25	19
The choice of bowl has no effect on the probability*	38	53
Select one card from each bowl at the same time.	22	18

[a] The response rate was .96 for grade 7 and .86 for grade 11.
*Indicates correct response.

items in table 3.2 posed the situation that Patty had rolled five 5s in a row and asked the probability of her getting a 5 on her next roll. Only about 40 percent of the 7th-grade students and 65 percent of the 11th-grade students responded correctly with 1/6. About a fourth of the 7th-grade students and a fifth of the 11th-grade students chose 5/6. There are a number of plausible explanations for this choice, including the simple use of the number given in the item, 5, and the number of outcomes possible with a die. It is possible these students believed, after five 5s in a row, that the die was not fair or that previous outcomes influence later ones and therefore that a 5 was a very strong possibility for the next roll also.

The responses to the item shown in table 3.5 give further evidence of confusion among some students about the independence of events. Approximately half the 7th-grade and 11th-grade students answered this exercise correctly. However, the number of students who responded that after four successive tails the coin was due to land heads up indicates that many students do not really believe in the independence of these events. This incorrect response reflects a common misconception known as "the law of small numbers," that is, the belief that in probability experiments the same pattern should emerge in the short run as one would expect in the long run. In the coin toss, according to this fallacy, the coin should begin to land heads up to balance the number of heads and the number of tails. Again, students' intuitive notions of probability seem to interfere with their applying formal knowledge of probability.

DISCRETE MATHEMATICS

Table 3.5
Independence of Events

Item	Percent Responding[a]	
	Grade 7	Grade 11
If a fair coin is tossed, the probability it will land tails up is 1/2. In four successive tosses the coin lands tails up each time. What happens when it is tossed a fifth time?		
It will most likely land heads up.	24	22
It is more likely to land heads up than tails up.	14	11
It is more likely to land tails up than heads up.	15	8
It is equally likely to land tails up or heads up.*	47	56

[a] The response rate was .98 for grade 7 and .94 for grade 11.
* Indicates correct response.

Permutations and Combinations

Permutations and combinations are closely related to probability, since students often need to correctly determine the number of arrangements or orderings in order to correctly determine the probability of an event. Most studies of discrete mathematics include specific techniques for determining the number of arrangements or orderings through the use of permutations and combinations. Many simple ordering or arrangement problems can be answered by using simple counting strategies. The formulas for finding permutations or combinations are simply extensions of these counting strategies.

The fourth assessment included three items relating to permutations and combinations at the 7th-grade level and four at the 11th-grade level. There were no items relating to these topics for 3rd graders. Though the number of items on the assessment related to these topics is minimal, some patterns emerge from the data.

Students in grades 7 and 11 were given a simple application item involving combinations in which they had to determine how many different skirt-blouse outfits could be made from a given number of skirts and blouses. The item could have been answered using a counting strategy or might have been recognized as being an application of multiplication. Students in grade 7 did rather poorly on this item; only 38 percent of them answered correctly as compared to 63 percent of the 11th-grade students. On a similar item from the second national assessment, 68 percent of the 13-year-olds answered correctly. One apparent difference between the two items was the inclusion of a picture with the item on the earlier assessment. It may be that students at this age can better understand the question and reason through the situation to a solution if they are given visual cues.

However, considering the nature of this problem and the assumption that students at these grade levels are familiar with multiplication, a much higher level of performance would be expected.

In general, 7th- and 11th-grade students had great difficulty with items related to permutations. On the item shown in table 3.6, two incorrect foils were selected more frequently than the correct one by 7th-grade students and one of them was selected more frequently than the correct one by 11th-grade students. It may be that students simply misunderstood the problem, since 25 would be the correct answer if the same number could be selected for both digits. However, a high percentage of students responded with 10, suggesting that they simply multiplied the two numbers that appeared in the problem, a technique used frequently by students when answering the probability items, or that they added 5 + 5. Very few students indicated that they did not know the answer to this item, suggesting that they believed that they could select the correct answer by some means. Furthermore, students may have intuitive ideas about counting problems that interfere with their application of formal knowledge, or they may lack enough formal knowledge to have a combinatorial view of multiplication.

Table 3.6
Permutations

Item	Percent Responding[a]	
	Grade 7	Grade 11
⬚1⬚ ⬚2⬚ ⬚3⬚ ⬚4⬚ ⬚5⬚		
In a game you are given these 5 cards. A rule says you must select 2 cards and form a 2-digit number such as		
⬚5⬚ ⬚2⬚		
How many different 2-digit numbers can you form including the one above?		
10	23	14
15	9	5
20*	20	28
25	25	34
30	12	11
I don't know.	11	8

[a] The response rate was .85 for grade 7 and .93 for grade 11.
*Indicates correct response.

Students in grade 11 were also given an item that asked the number of different ways that six different objects could be arranged on a shelf. About half selected the response 36; only about a fourth selected the correct response. Although this item was very different from the one in table 3.6, in both items the foil most frequently selected, although incorrect, was that which could be obtained by multiplying the number of objects by itself. Furthermore, on an item asking the number of different ways a committee of three students could be selected from a class of ten students, almost two-thirds of the 11th-grade students incorrectly chose 30, the number obtained by multiplying the two numbers in the problem. There is no evidence that students actually calculated their answer in this way; however, the pattern that emerges suggests this error.

4

DATA ORGANIZATION AND INTERPRETATION

Catherine A. Brown   *Edward A. Silver*

THE schools are giving increased attention at all grade levels to the organization, analysis, and interpretation of data: students learn to read and interpret graphs and to find and use measures of central tendency and variability in different problem settings. As early as the elementary grades, teachers introduce students to simple descriptive statistical methods and various graphical means of presenting data. The fourth assessment included sixteen items at grade 3, twenty-three items at grade 7, and twenty-three items at grade 11 that dealt with these topics; the items were distributed fairly evenly among the topics.

Unfortunately, many items on descriptive statistics appeared late in the item blocks; thus on many of them the response rates fell below 50 percent for 11th-grade students and below 75 percent for 7th-grade students. Although this report includes data from some of these items because they are essential for comparing results across grade levels, the reader should take account of the response rates when considering the results. Most of the items dealing with graphs, tables, and charts had acceptable response rates.

HIGHLIGHTS

• Students at all three grade levels performed poorly on items related to measures of central tendency and variability.

• Most students in the 7th and 11th grades appeared not to understand technical statistical terms such as *mean, median, mode*, and *range*. However, there is evidence that they could compute the *mean* when asked for the *average*.

• Students at all grade levels generally could read and compare data in table and graph form.

• Most students had difficulty when asked to use data presented in table or graph form to do problem-solving tasks or when the data presentation or the task were somewhat different from standard textbook exercises.

DESCRIPTIVE STATISTICS

More than half of the items on descriptive statistics dealt with measures of central tendency and variability and included such technical terms as *mean, median, mode,* and *range.* Knowledge of these topics was assessed primarily at the 7th- and 11th-grade levels. Few elementary school students have much experience with measures of central tendency and variability in their mathematics classes, however, these topics are often included in the middle, junior high, and high school mathematics curricula. The fourth assessment, thus, reflects the present situation in school mathematics.

Students in grade 11 did somewhat better than students in grade 7 on items involving mean, median, and mode. However, these items were difficult for students at both grade levels, as shown by the results presented in table 4.1: fewer than half of students at either grade level answered these items correctly. In another item, 11th-grade students were given the data in table 4.1 and were asked to select the statement that best described the change that would occur in the statistics if 32 inches of snow fell during January of 1985. Only about 40 percent of the students selected the correct statement.

Table 4.1
Mean, Median, and Mode

Inches of Snow in January	
Year	Inches of Snow
1970	15
1971	16
1972	17
1973	15
1974	15
1975	16
1976	16
1977	18
1978	15
1979	17
1980	15
1981	17
1982	16
1983	17
1984	15

	Percent Correct [Response Rate]	
Item	Grade 7	Grade 11
A. What is the mode?	26 [.65]	40 [.41]
B. What is the median?	38 [.65]	47 [.41]
C. What is the mean?	40 [.66]	41 [.72]

The data presented in table 4.2 indicate that older students did much better on an item that used the term *average* than on the similar item in table

Table 4.2
Averages

	Percent Correct [Response Rate]	
Item	Grade 7	Grade 11
Here are the ages of six children: 13, 10, 8, 5, 3, 3 What is the average age of these children?	46 [.94]	72 [.98]

4.1 that used the term *mean*. (Very few 3rd-grade students successfully answered this item, and almost half answered that they did not know the average; these results indicate their lack of experience with these topics.) The results suggest that older students are more familiar with the term *average* than the term *mean* and can compute this statistic when they understand what the question is asking. There is evidence, however, that although many 7th-grade and 11th-grade students are able to calculate averages when asked to do so, the depth of their understanding of the concept of average is rather shallow. For example, when asked a question about the size of the average of two given numbers relative to those two numbers, only about 40 percent of the 7th-grade students and about 50 percent of the 11th-grade students correctly answered that the average must be halfway between the two numbers.

Responses to item A in table 4.3 suggest that most 11th-grade students know how to compute a weighted average. Although 30 percent of the students incorrectly picked the second foil, it seems likely that it was chosen by many through an inattention to detail; many students may have thought they were selecting the correct (i.e., third) answer instead of the incorrect one in which multiplication signs were substituted for addition signs.

Item B, a fairly complex problem, was another for which students had to find the weighted average. Very few students at either grade level answered this item correctly, although as stated above, many of them seem to know how to calculate weighted averages. The number of students who selected an average cost that was either less than or greater than both of the given costs further indicates a shallow understanding of the concept of average.

Table 4.4 contains samples of items that asked 11th-grade students to use statistical information in word-problem settings. Performance by students on items A and B was very poor. The percentages suggest that students were guessing when answering these items, indicating an inability to apply their knowledge of average. Their difficulty with item B cannot be totally accounted for by a lack of understanding of the term *range*, since on two other items related to variability, more than half of the 11th-grade students correctly identified the range of a set of numbers.

DATA ORGANIZATION AND INTERPRETATION

Table 4.3
Weighted Averages

	Item	Percent Responding[a]	
		Grade 7	Grade 11
A.	Score A = 4 B = 3 C = 2 D = 1 F = 0 Frequency 8 7 0 3 5 Which of the following procedures will give the average grade for the test scores given above?		
	$\dfrac{4 + 3 + 2 + 1 + 1}{18}$		19
	$\dfrac{(8 \times 4) \times (3 \times 7) \times (2 \times 0) \times (1 \times 3) \times (0 \times 5)}{23}$		30
	$\dfrac{(8 \times 4) + (7 \times 3) + (0 \times 2) + (3 \times 1) + (5 \times 0)}{23}$*		43
	$\dfrac{(4 \times 8) + (3 \times 7) + (3 \times 1)}{18}$		9
B.	Louise bought some packages of fudge. 2 pounds of vanilla for 90¢ per pound 3 pounds of chocolate for $1.60 per pound What was the average cost per pound for this fudge?		
	$ .50	17	24
	$1.25	14	22
	$1.32*	12	20
	$2.50	25	11
	$3.30	6	3
	$6.60	16	11
	I don't know	14	9

[a] The response rate on item A for grade 11 was .84. On item B the response rate was .50 for grade 7 and .58 for grade 11.
* Indicates correct answer.

CHARTS, GRAPHS, AND TABLES

Reading and interpreting information in graphs, charts, and tables are important information-processing skills. People regularly see information in the form of graphs, charts, and tables in newspapers and magazines, on television, and in reports from various sources. Information is displayed in various formats, such as bar graphs, line graphs, circle graphs, tables, and charts.

In the fourth assessment, students at each grade level were given approximately fifteen items that involved the various forms of data presentation. Several types of questions were asked about the data displayed in these items. Some asked students to read information directly from a display, that is, to locate data on a chart, graph, or table and report it. More complicated questions asked students to compare two or more entries in the

Table 4.4
Statistical Word Problems (Grade 11)

Item	Percent Responding[a]
A. Edith has an average (mean) score of 80 on five tests. What score does she need on the next test to raise her average to 81?	
81	14
82	31
85	31
86*	24
B. Bill made the lowest score on the test. He only got 29 points. The teacher said the class mean was 65 and the range was 51. Jane made the highest score on the test. What score did Jane make?	
51	12
65	29
80*	43
94	16

[a] Response rate was .48 for item A and .68 for item B.
* Indicates correct answer.

display or to use data from the display to solve a problem that might require some computation. Finally, there were questions that required students to interpolate or extrapolate from known data. As one might expect, problem-solving items and items involving interpolation or extrapolation were the most difficult for students.

Table 4.5 shows the results for two sets of items presented to students. One set of items presented data in graph form and the other gave the same data in table form; students in grade 11 were given only the graph items. Most students at each grade level did well on the items (IA and IIA) that required only a direct reading of the bar graph or the table. Except for 3rd-grade students, who may have had difficulty with the addition, items IC and IIC, which required students to read data from the table and then add, were also successfully answered by most students. Students had considerable difficulty answering items IB and IIB, which required them to determine relationships among the quantities of three fruits and then draw a conclusion.

It is interesting to note the small differences in performance on the two sets of items. One would expect that using numerical data from the tables would have been easier than reading the graphs and thus would expect somewhat higher performance on items IIA, B, and C than on items IA, B, and C. For items IC and IIC, this expectation was met for 3rd-grade students, but the results for 7th-grade students were identical. At both lower grade levels, performance was only slightly different for items IA and IIA. Items

DATA ORGANIZATION AND INTERPRETATION

Table 4.5
Graphs and Tables

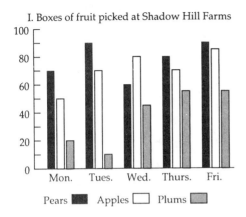

I. Boxes of fruit picked at Shadow Hill Farms

II. Boxes of fruit picked at Shadow Hill Farms

Day	Number of boxes		
	Pears	Apples	Plums
Mon.	70	50	20
Tues.	90	70	10
Wed.	60	80	45
Thur.	80	70	55
Fri.	90	85	55

			Percent Correct [Response Rate]		
Item			Grade 3	Grade 7	Grade 11
A. How many boxes of pears were picked on Thursday?		Graph	67 [.86]	87 [.98]	91 [.99]
		Table	70 [.96]	91 [.98]	
B. On which day were more boxes of apples picked than either boxes of pears or boxes of plums?		Graph	29 [.82]	64 [.98]	74 [.99]
		Table	34 [.96]	76 [.98]	
C. How many boxes of pears, apples, and plums were picked on Tuesday?		Graph	44 [.79]	86 [.98]	95 [.99]
		Table	58 [.95]	86 [.98]	

IB and IIB were the most difficult for students, and differences in performance on the table and graph forms were greater than on the other items.

Additional evidence that students can read and interpret information displayed in tables and graphs is given in the level of success—better than 60 percent—on most other items involving this skill. It should be noted that 3rd-grade students had some difficulty with an item that asked them to determine where the top of a bar should be placed in a bar graph. This suggests that although younger students can read and use graphs, they have trouble constructing them. (Items of this sort were not presented to the older students.)

A few items involved interpolating or extrapolating information from graphs. Students in grades 7 and 11 were given the items shown in table 4.6; the 11th-grade students were considerably more successful on these items

Table 4.6
Interpolation from a Graph

This graph shows how far a typical car travels after the brakes are applied.

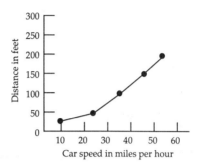

	Percent Correct [Response Rate]	
Item	Grade 7	Grade 11
A. A car is traveling at 30 miles per hour. About how far will it travel after applying the brakes?	43 [.98]	65 [.99]
B. A car traveling down the street stopped 100 feet after the brakes were applied. About how fast was the car traveling?	61 [.98]	83 [.99]
C. A car is traveling 55 per miles per hour. About how far will it travel after applying the brakes?	41 [.98]	70 [.99]

than the 7th-grade students. The results on these and other items suggest that more than two-thirds of 11th-grade students can successfully interpolate or extrapolate information from graphs but fewer than two-thirds of 7th-grade students can do so.

5

MEASUREMENT

Mary M. Lindquist   Vicky L. Kouba

THE discussion of the approximately 50 measurement items is organized into three categories. The first category focuses on the unit of measure, both customary and metric units. The second category examines the ability of students to use measuring instruments. The third category contains items related to perimeter, area, and volume.

Items from these categories were administered to each of the grades. Only two items were common to all three grade levels. Most of the items were given at two grade levels, however, enabling change in performance across grades to be observed.

The set of items includes some, but not all, of the measurement concepts and skills emphasized in schools. There were items that required knowing the basic conversion equivalences, reading simple measurement instruments, and finding area and perimeter of rectangles. None of the items required telling time on nondigital clocks, converting to different units of measure, or finding areas of triangles or parallelograms.

HIGHLIGHTS

• Students at all grade levels were able to read a variety of simple measurement instruments.

• There is an improvement of about ten percentage points from the 7th to the 11th grade on items involving units of measurement. A similar growth was found for area and volume items.

• Neither 7th- nor 11th-grade students have developed a strong conceptual understanding of area.

• Although three-fourths of the 11th-grade students can find the area of a rectangle, fewer than half can use this skill in related problems.

UNITS OF MEASUREMENT

This category included sixteen items that assessed students' abilities to recognize the relationship between the size of the unit and the number of units, select an appropriate unit, estimate the size of objects, and state the equivalence relation between units. There were no items that required conversion between units within systems, although two items required knowledge of the relative size of meters and yards.

There are many concepts associated with the role of the unit in measuring that are crucial to understanding measurement. One of these concepts is concerned with the relationship between the unit and the number in a measurement. Although this concept is not stressed in the curriculum, students with appropriate experiences realize that it takes more smaller units than larger units to measure an object. Two items were administered to ascertain 3rd-grade students' understanding of this concept. Almost two-thirds of the students at this grade level could tell that it takes more of a smaller unit than of a larger unit to fill a box. In the other item, the units were not pictured. The students were told that the measurements of an object were made with different units, and the measurements were given in terms of the number of units. For example, Sam reported the length of an object to be 8 of his units, and Sue reported that it was 6 of her units. Both 3rd-grade and 7th-grade students had difficulty with this item. Over half named the person who used the most units as the one with the largest units. That is, they focused only on the number. This may have been because of the testing situation and question, but it also indicates that this concept is not completely understood by students at these grades.

Third-grade and 7th-grade students were given a pair of items similar to those in table 5.1 in which they had to select the appropriate metric unit. The centimeter was the appropriate unit for the pen, and the meter was the appropriate unit for the length of the house. Both groups of students appeared to be more familiar with the smaller unit than with the larger one.

Students in grade 3 were asked to choose an estimate of an object's length in inches. About 40 percent of the students chose the best estimate; another

Table 5.1
Appropriate Units

Item	Percent Correct [Response Rate]	
	Grade 3	Grade 7
Select appropriate unit to measure the length of a pen.	60 [.94]	87 [.97]
Select appropriate unit to measure the length of a house.	35 [.48]	67 [.85]

40 percent chose a length that was a fairly good estimate. It appears that over three-fourths of the students had some knowledge of estimating in inches.

When 3rd-grade students were asked to choose an estimate in metric units of the weight of a familiar object, almost half indicated that they did not know. By 7th grade, about two-thirds of the students chose the appropriate metric unit to weigh a large object. Although these items were different, the results indicate improvement from one grade level to the next. Another item, administered to all three grades, asked for the largest of several metric units of length. The performance of 3rd-grade students was about at chance, whereas 62 percent of the 7th-grade students and 72 percent of the 11th-grade students chose the largest unit. This improvement of about ten percentage points from 7th to 11th grade held almost without exception for items given to both grades.

Three items were given to both 7th- and 11th-grade students that required them to recall the equivalence of metric units (e.g., 1000 meters equals 1 kilometer). Over 40 percent of the 7th-grade students and over 50 percent of the 11th-grade students responded correctly to such items. In contrast, fewer than 10 percent of the 7th-grade students and fewer than 30 percent of the 11th-grade students could state the equivalence between customary area units. One must interpret these results with caution, since the metric items were multiple choice with only three foils, whereas the customary item was open-ended.

The only set of items that allowed a direct comparison of students' understanding of metric units and of standard units was a pair of parallel items asking them to estimate the height of a familiar object in feet and in meters. About 70 percent and 80 percent of the students in grades 7 and 11, respectively, chose the best estimate when it was given in feet. As shown in table 5.2, only half these students chose the best estimate when it was given in meters.

Table 5.2
Length Estimation

	Percent Correct [Response Rate]	
Item	Grade 7	Grade 11
Select closest measurement in feet to height of a tall man.	72 [.97]	83 [.99]
Select closest measurement in meters to height of a tall man.	34 [.81]	44 [.96]

MEASUREMENT INSTRUMENTS

Six items assessed students' ability to read or use measurement instruments, and four items assessed their knowledge of precision of measurement. The instruments pictured on the assessment were rulers, scales, a thermometer, and a digital watch. No items required the actual handling of an instrument or telling of time shown on a nondigital clock.

Generally, students were able to read the instruments presented. Over three-fourths of the 3rd-grade students could read a scale showing a weight of 78 pounds. They were able to count the tick marks between 75 and 80 to arrive at this answer. About 65 percent and 90 percent of the 3rd-grade and 7th-grade students, respectively, could read the thermometer, which showed a temperature of 10 degrees below zero. The lower result for 3rd-grade students on the temperature item than on the weight item may be due to the negative sign by the numeral 10 on the thermometer. About one-fourth of the 3rd-grade students chose "10 above zero."

The 3rd-grade students were shown a digital watch reading 7:47 and were asked to choose the number of minutes until 8:00. Over half selected the correct number, an indication of their understanding of a digital watch and the number of minutes in an hour.

About three-fourths of the 7th-grade students could read a ruler to the nearest fourth of an inch. The ruler was aligned with the object, and the object ended very near a fourth-inch mark.

An item for 11th-grade students pictured a scale holding four identical objects weighing about 30 pounds all together. When asked to choose the approximate weight of each object, about three-fourths of the students answered correctly. The others chose the weight shown on the scale, probably not pausing to read the question carefully.

The previous discussion indicates that about three-fourths of the students can read these instruments. It must be noted, however, that these items represented some of the simplest tasks involved in reading instruments. There were no instruments calibrated in units other than one or that could not be read directly. For example, thermometers are often calibrated in increments of two. There were no items requiring reading of a nondigital clock, no items in which a judgment had to be made about rounding to the nearest fraction of an inch, and no items requiring a protractor. On past assessments, such items have been difficult for students.

The item shown in table 5.3 sheds some light on students' ability to read a ruler in a more complex situation than in the previous discussion. In this case, the segment is not aligned at the beginning of the ruler. Few of the 3rd-grade students and about half of the 7th-grade students chose the correct length. Although almost three times as many 3rd-grade students as 7th-grade students chose the endpoint, about a third of the students at each grade level chose 6 cm. One may speculate that those who chose 6 cm were counting numbers rather than units.

Two precision items were given to 7th-grade students, but the response rates were so low that it is difficult to interpret the results. About half the

Table 5.3
Rulers

	Percent Responding[a]	
Item	Grade 3	Grade 7
How long is this line segment?[b]		
3 cm	4	1
5 cm*	14	49
6 cm	31	37
8 cm	30	9
11 cm	6	2
I don't know.	15	2

[a] The response rate was .80 for grade 3 and .97 for grade 7.
[b] An actual centimeter ruler was pictured.
*Indicates correct response.

11th-grade students had some knowledge of the precision of measurements of length. Although more than half chose the correct answer for the item in table 5.4, students may not have been fully aware that they were choosing a more precise measurement, since the correct answer is also the one that is nearest to 6 m or that can be rounded to 6 m. In the other items, about half the 11th-grade students chose the most precise measurement from a list of length measurements, told the amount of possible error in a given length measurement, and told how error in measuring length could be reduced.

Table 5.4
Length Precision—Grade 11

Item	Percent Responding[a]
The length of an object was measured and found to be 6 meters *to the nearest meter*. Which of these could have been the length if the object was measured with greater precision?	
5.3 m	4
6.8 m	17
6.53 m	11
5.6 m*	68

[a] The response rate was .96.
*Indicates correct response.

PERIMETER, AREA, AND VOLUME

Eighteen items assessed students' knowledge of perimeter, area, and volume. Because 3rd-grade students are not expected to have mastered these concepts and skills, they received only four items. Eleven of the items were given at grade 7 and 14 items were given at grade 11. One item was given to students at each of the grade levels and eight of the items were to both 7th- and 11th-grade students, so that growth in performance could be analyzed.

Performance on perimeter and area items improved from grade to grade, as shown in table 5.5. Third-grade students have little knowledge of these topics; many of them added the two numbers shown or readily admitted that they did not know by choosing the "I don't know" option. Although 7th-grade students showed a better grasp of these topics, their performance

Table 5.5
Area and Perimeter

	Percent Correct [Response Rate]	
Item	Grade 3	Grade 7
What is the perimeter of this rectangle?	17 [.62]	46 [.93]
What is the distance around a 4-by-7 rectangle?	15 [.59]	37 [.70]
What is the area of this rectangle?	20 [.70]	56 [.96]
What is the area of this rectangle?	5 [.63]	46 [.86]

MEASUREMENT

was less than 50 percent on most items. By grade 7, many students were confusing area and perimeter. The most common error made on the area items was choosing the perimeter, and vice versa. This confusion was not entirely eliminated by grade 11.

There is evidence that the confusion between perimeter and area is not the only misconception that students have about area. Both 7th-grade and 11th-grade students received an item that required the knowledge that the sum of the areas of the parts of a rectangle equals the area of the whole. About one-fourth of the students said that the sum of the area of the separated parts could not be determined. Another item asked for an estimate of the area of an irregular figure placed on a grid, such as the one shown in table 5.6. About 40 percent of the 7th-grade students and about 55 percent of the 11th-grade students chose the best of the given estimates. About one-fourth of the students at both grade levels chose 36, the number of squares that were at least partially shaded.

Table 5.6
Estimating Area

	Percent Responding[a]	
Item	Grade 7	Grade 11

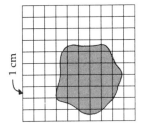

About how many square cm is the area of the shaded region?[b]		
16	19	13
26*	40	55
36	28	24
52	2	1
I don't know.	10	6

[a] The response rate was .98 for grade 7 and .99 for grade 11.
[b] Actual square centimeter paper was used.
*Indicates correct response.

Given one side of a square, little more than 10 percent of the 7th-grade students and about 45 percent of the 11th-grade students could determine the area. The same percentage of 11th-grade students could find the area of a square given its perimeter. The students perhaps performed less well on these items because they did not understand that a square is a special case of a rectangle or that it is necessary to first determine both dimensions.

Table 5.7
Area and Volume

	Percent Correct [Response Rate]	
Item	Grade 7	Grade 11
What is the volume of this rectangular solid?	33 [.93]	58 [.93]
What are the dimensions of this rectangular solid?	43 [.83]	68 [.97]
What is the area of this rectangle?	46 [.89]	70 [.99]
What is the area of this rectangle?	56 [.96]	

Several problems given to the 11th-grade students required more steps. Fewer than 30 percent of the students could find the area of a figure composed of two rectangles. In another problem, the students were told that a photograph had been enlarged by doubling its dimensions. About 45 percent of the students chose the cost of the enlarged photograph given the price per square inch.

Some items asked students in grade 7 and 11 to find the volume or dimensions of rectangular solids. Performance on these items was about ten percentage points below corresponding area items. Note in table 5.7 the constant amount of improvement from grade 7 to grade 11 as well as the constant difference between items in which the grid is shown and items in which only the dimensions are given.

Two items asked the 11th-grade students to estimate the circumference of a circle given the radius in one item and the side of the circumscribed square in the other item. The correct-response level for each item was little better than chance. When given the radius, about one-third of the students selected the length of the diameter instead of the circumference. When given the side of the circumscribed square, almost half chose the perimeter of the square, not realizing that the circumference would be less than the perimeter. In neither case was it necessary to know the formula for the circumference, since the picture could have been used in making a reasonable estimate. The low performance on these items probably stems from students' lack of experience in making estimates of the lengths of curves.

6

GEOMETRY

Mary M. Lindquist   *Vicky L. Kouba*

THIS assessment included forty-eight geometry items—six were given to 3rd-grade students, twenty-nine were given to 7th-grade students, and forty-three were given to 11th-grade students. For this discussion, the items are sorted into four categories: identification of figures, properties of figures, visualization tasks, and applications.

Each category contains enough items to give a view of the performance of the 7th-grade and 11th-grade students. The results of the six items given to 3rd-grade students are mentioned in the appropriate sections, but few generalizations can be made from such a limited number of items.

HIGHLIGHTS

• Students performed best on items that require the identification of common geometric figures. Their next best performance was on visualization tasks, followed by items requiring knowledge of geometric properties. Application items were the most difficult.

• Eleventh-grade students who had not taken high school geometry performed at approximately the same level as 7th-grade students.

• Few differences in performance were found between 11th-grade students who had and had not taken geometry on items that depended mainly on visualization, some differences were found on items involving geometric terms, and great differences were found on items requiring a knowledge of geometric properties or applications.

IDENTIFICATION OF FIGURES

Unfortunately, much of the geometry that students study prior to high school focuses only on identification of figures and geometric terms. No items in this category were administered to 3rd-grade students. The older students did well on many of the items in this category, but their performance was not uniform, as shown by the results in table 6.1.

Table 6.1
Identification of Figures

	Percent Correct [Response Rate]			
Figure	Grade 7	Grade 11 (All)	Grade 11[a] (No Geometry)	Grade 11[a] (Geometry)
A. Parallel line	90 [.98]	96 [.99]	93	99
B. Perpendicular lines	33 [.86]	63 [.99]	25	88
C. Sphere	67 [.67]	76 [.99]	57	87
D. Diameter	74 [.77]	81 [.98]	81	95
E. Radius	44 [.75]	66 [.98]	41	81
F. Endpoints of arc	47 [.72]	72 [.97]	49	87

[a] Response rates for students who had and had not taken geometry were approximately the same as the response rate for all students in grade 11.

Most students were able to identify parallel lines, but only about one-third of the 7th-grade students and two-thirds of the 11th-grade students could identify perpendicular lines. A similar, but not quite as dramatic, difference may be noted in students' identification of *diameter* and *radius*. Students knew the terms used more commonly in everyday life *(parallel* and *diameter)* than those used in more technical ways *(perpendicular* and *radius)*. This contrast was much less pronounced for students who had taken geometry than it was for those who had not taken geometry.

Table 6.1 also shows the similarity between the performance of the students in grade 7 and those in grade 11 who have not taken geometry. It does not appear that these 11th-grade students have had any additional opportunities in school to learn any geometric terms. They did slightly better than the 7th-grade students on the everyday terms but not as well on the more technical terms.

One item, which was given only to students in grade 7, required the identification of a quadrilateral that was *not* a parallelogram. Over half of the students chose the correct response (a trapezoid), but one-fourth chose a square. Very few of the students chose the foil that was a typical picture of a parallelogram or the foil that was a rectangle. Their performance on this item indicates that students of this age do not have a fully developed concept of the inclusion relationships among quadrilaterals. That is, they do not understand that squares are rectangles, that rectangles are parallelograms, and thus, that squares are parallelograms.

PROPERTIES OF FIGURES

Approximately twenty items were administered that required identification of the properties of figures or simple applications of the properties of triangles, circles, angles, and squares. Two items involving symmetry were also given. The items are discussed as they relate to various figures, but some commonalities across performance are noted.

Triangles. The five items described in table 6.2 require some knowledge about the properties of a triangle. Several different patterns in the responses appear in this set of items as well as the other figures. First, performance of those 11th-grade students who had not taken geometry was at the same level as 7th-grade students.

Table 6.2
Properties of Triangles

	Percent Correct [Response Rate]			
Item Description	Grade 7	Grade 11 (All)	Grade 11[a] (No Geometry)	Grade 11[a] (Geometry)
A. Choose the sets of 3 line segments that *cannot* make a triangle.	67 [.97]	—	—	—
B. Choose the set of 3 numbers that *cannot* be sides of a triangle.	9 [.50]	24 [.58]	8	33
C. Choose the set of 3 numbers that *cannot* be angles of a triangle.	18 [.87]	51 [.93]	17	71
D. Choose a true statement about angles of a triangle given that two sides are equal. [Stem and statement expressed in words.]	38 [.75]	58 [.90]	42	70
E. Same as item D except expressed in symbols.	—	59 [.64]	35	71

[a] Response rates for students who had and had not taken geometry were approximately the same as the response rate for all students in grade 11.

Second, if an item could be solved visually, then performance was better than if it required more abstract thinking. Given the three line segments in item A in table 6.2, over two-thirds of the 7th-grade students could mentally visualize or tell in some way which set would or would not form a triangle. Item B can be solved in at least two ways: visualizing the lengths of each segment and then continuing as in item A or knowing that the sum of the lengths of any two sides must be greater than the length of the third side. Evidently, neither strategy was in the repertoire of 7th-grade students or of 11th-grade students, even those who had taken geometry. In contrast, two-thirds of those students who had geometry did know that the sum of the angles of a triangle is 180. They were successful in choosing the triple of numbers that could not be the measures of angles of a triangle in item C.

The 11th-grade students handled items that were stated verbally at the same level as they did those that were stated more symbolically. The symbols used in item E were not overly complicated but were peculiar to geometry. The use of symbols did decrease the number of words in the problem and, thus, it may have been processed more quickly.

Another set of items, not reported in table 6.2, was given only to 11th-grade students. Given the size of one of the angles of a triangle, the students were asked if the triangle could be, could not be, or must be certain a type of triangle. Correctly answering this set of items depended on knowing about the angles of the different types of triangles as well as the sum of the angles of a triangle. As might be expected, those students who had taken geometry did much better than those who had not. Generally, about two-thirds of the students who had taken geometry answered correctly in contrast to fewer than one-third of those who had not taken it.

Circles. Three items that involved properties of circles are described in table 6.3. Items A and B both required students to know or to visualize that the points on a circle are all the same distance from the center. For item A the students were given a description of the situation accompanied by a diagram of three points placed around a reference point. For item B, they were given only a verbal description. Although item A had more of a real-life setting and item B was worded more mathematically, the results are almost identical. About one-third of the 7th-grade students and two-thirds of the 11th-grade students chose the circle. In item A the most commonly

Table 6.3
Properties of Circles

	Percent Correct [Response Rate]			
Item Description	Grade 7	Grade 11 (All)	Grade 11[a] (No Geometry)	Grade 11[a] (Geometry)
A. Choose the figure formed by people standing the same distance from a given person.	35 [.91]	64 [.92]	41	79
B. Choose the figure whose points are same distance from a given point.	35 [.71]	66 [.92]	42	81
C. Find the radius of a circle inscribed in a square.	28 [.76]	58 [.86]	33	72

[a]Response rates for students who had and had not taken geometry were approximately the same as the response rate for all students in grade 11.

chosen distracter was the triangle. This may have been because the people shown in the diagram formed a triangle and the students did not read the whole problem carefully. In item B the most commonly chosen distracter was a square in which all the vertices were the same distance from the given point; the eye tends to focus on these four points. Throughout the geometry items, students were often misled by visual clues, such as those described here. The 3rd-grade students were also given item A, but the results were at the chance level, so they are not discussed in detail.

Only slightly fewer students could find the length of the radius of a circle inscribed in a square whose side was given (item C, table 6.3) than could identify the radius of a circle. Again, it is interesting to note that about three-fourths of those students who had taken geometry responded correctly to this item, which is about the same percentage of geometry students who responded correctly to items about triangles.

One other item on circles was given, but the results must be viewed with caution, since the response rate was only about 50 percent. The results are included here because they raise some questions that teachers might use in talking with students. In the item, a circle was presented with an inscribed triangle, one side of which was a diameter. The 11th-grade students were asked to tell which of three statements about the triangle were true. About one-third of those who had not taken geometry and half of those who had knew that the inscribed triangle was a right triangle and that two sides were perpendicular. Almost half of the students who had not taken geometry responded that the two nonright angles must be equal even though they clearly were not equal in the picture. It would be interesting to have students give reasons for their choices or to consider the item if the right triangle was turned in the conventional position.

Angles. The results of five items dealing with vertical or supplemental angles are reported in table 6.4. About three-fourths of the students who had taken geometry answered correctly these items, but those who had not taken geometry did not perform well. Although a difference in performance may be expected, the concepts tested by these items are found in most junior high school programs.

As mentioned, the visual cue in item C (table 6.4) strongly influenced the students. Over three-fourths of all students responded correctly to this item. The influence of the visual cue also was evident in item E. Students who had not taken geometry could identify that vertical angles must be equal but failed to realize that the sum of the supplementary angles was 180.

Squares. One of the two items related to squares was given only to 3rd-grade students. About two-thirds of these students could tell the length of a side of the square given the length of an adjacent side. The other item was given to both 7th-grade and 11th-grade students, but the response rate for 7th-grade students was too low to consider the results as meaningful. Fewer

Table 6.4
Vertical and Supplemental Angles

Item Description	Percent Correct [Response Rate]			
	Grade 7	Grade 11 (All)	Grade 11[a] (No Geometry)	Grade 11[a] (Geometry)
A. Identify a pair of angles as supplemental.	—	58 [.98]	42	70
B. Identify that the combined measure of supplemental angles is 180.	—	56 [.71]	23	76
C. Find the measure of angle a.	75 [.97]	86 [.99]	71	95
D. Find the measure of angle b.	10 [.97]	49 [.99]	14	71
E. Choose the true statements about vertical and supplemental angles.	24 [.70]	56 [.91]	24	75

[a] Response rates for students who had and had not taken geometry were approximately the same as the response rate for all students in grade 11.

than half of the 11th-grade students could give the length of a diagonal of a square when given its area. This is not surprising if one considers the results of items on area discussed in chapter 5. What is discouraging is that only a little over half of those students who had taken geometry responded correctly to this item. Almost one-fourth of these students chose the "not enough information" option.

Symmetry. One of the two items involving symmetry was given only to 3rd-grade students. Most students were able to choose the picture that could not be folded so that the edges would match. The other item involving symmetry was given only to 11th-grade students. Although the majority chose a picture with a vertical line of symmetry, many also responded that a parallelogram had a line of symmetry parallel to one pair of opposite sides.

Some also chose a figure that had rotational, but not reflectional, symmetry. Fewer than one-fourth of 11th-grade students, whether or not they had taken geometry, chose the correct response. It is evident that 11th-grade students have some correct ideas about symmetry but also some misconceptions.

VISUALIZATION TASKS

Many items relied on visual cues, but one group of items could also be classified as more visual than geometric in nature. A few of the items involved simple geometric terms, but others used only everyday words. The performance on this group of items was fairly uniform; about half of the 7th-grade students and over two-thirds of the 11th-grade students responded correctly. Students who had taken high school geometry performed only slightly better than those who had not. Their performance varied greatly when geometric terms as well as visualization were involved.

In table 6.5, a pair of items is described that requires visualization of an object or scene from a different perspective. Students in all three grades were more successful with the item that required them to select a top view of a block of wood than with one that required them to visualize a scene from another person's perspective. Note that in both items, students who had geometry did somewhat better than those who had not, but the difference was not as marked as on most of the other geometric items. It is interesting to note that the 7th-grade students did not perform much better than the 3rd-grade students on item A, but they did considerably better on item B. The type of incorrect response pattern for these two grades accounts for this difference. On both items, about one-third of the 7th-grade students chose the foil that repeated one part of the picture. That is, in item A they chose the top drawn in perspective rather than a top view and in item B they chose the picture as drawn rather than the picture viewed from the perspective of

Table 6.5
Change of Perspective

	Percent Correct [Response Rate]				
Item Description	Grade 3	Grade 7	Grade 11 (All)	Grade 11[a] (No Geometry)	Grade 11[a] (Geometry)
A. Top view of a block	50 [.75]	53 [.63]	71 [.95]	60	78
B. View of scene between you and person facing you from his perspective.	20 [.95]	39 [.60]	54 [.83]	43	56

[a] Response rates for students who had and had not taken geometry were approximately the same as the response rate for all students in grade 11.

GEOMETRY

the other person. About the same percentage of the 3rd-grade students also chose this incorrect option for item A, but almost twice as many of them were unable to put themselves in the place of another person in order to answer item B correctly. Although this response is not surprising for this age group, it should serve as a reminder to all of us who stand in front of children: their view is not the same as ours when we are looking at them.

Two items on transformations were given only to 7th- and 11th-grade students. About two-thirds of the students at each grade level chose the correct image of a reflection. About half of the 7th-grade students and three-fourths of the 11th-grade students chose the correct image of a turn. The difference in performance of the 7th-grade students on the two items might be the result of item's placement in the testing block. The turn item came late in the 7th-grade block, and about one-third did not respond to it.

The other items concerned the relationship between two- and three-dimensional figures. One item presented three networks consisting of five squares joined in different configurations. The 11th-grade students were asked to choose the networks that could be folded into cube without a top. About 45 percent of the students chose both of the networks, and 44 percent chose one of the two networks that could be folded into an open cube. Again, there was a difference in the performance of those who had and who had not taken geometry, but it was not as marked as in nonvisual items. More of the students who had not had geometry selected only one of the options.

Two items in table 6.6 deal with a cylinder and its two-dimensional components. Item A was easier than item B for each group of the 11th-grade students. It would be interesting to investigate whether the square, the terms, or the required processing made item B more difficult than item A.

APPLICATIONS

Seven items could be classified as applications. One of these was given only to 3rd-grade students: they were asked to count the number of triangles shown in a larger triangle. About two-thirds of the students counted the small, inside triangles but not the overlapping triangles or the outside triangle. Unless children have been exposed previously to this type of puzzle problem, it is not surprising that they would count only the most obvious triangles.

A rather simple problem, shown in table 6.7 was given to both 7th- and 11th-grade students. As in the items discussed previously, the visual clue might have been misleading. The students might have focused on the arc length of an angle rather than on the angle size. This result indicates that the *concept* of an angle is not well developed even though the *computation* needed to find its size is well developed by this time. Also note that as in previous items, about three-fourths of those students who had taken geometry responded correctly.

Table 6.6
Cylinders

	Percent Correct [Response Rate]		
Item Description	Grade 11 (All)	Grade 11[a] (No Geometry)	Grade 11[a] (Geometry)
A. If this circular, open-ended paper cylinder is cut along the dotted line and flattened out, it will always form what figure?	75 [.97]	60	78
B. Side Top — A geometry solid is viewed from the side and from the top. Those views are shown above. What could the solid be?	61 [.95]	43	72

[a] Response rates for students who had and had not taken geometry were approximately the same as the response rate for all students in grade 11.

Three other problems involved the Pythagorean relationship. The results of two of these items are shown in table 6.8 on page 54. The third item, given only to 11th-grade students, was more symbolic but had similar results. In general, all students did not perform well on these items. On item A, the most common response was one that might have involved counting using a unit different from the one specified. That is, the students possibly used the slanted distance between two adjacent horizontal lines as the unit, or they chose an answer that matched the number shown on the axis. Even without knowing the Pythagorean relationship, the students could have selected a more reasonable answer by realizing that the hypotenuse had to be longer than the sides. Although item B is slightly more complicated, the

GEOMETRY

Table 6.7
Application Angles

	Percent Correct [Response Rate]			
Item Description	Grade 7	Grade 11 (All)	Grade 11[a] (No Geometry)	Grade 11[a] (Geometry)
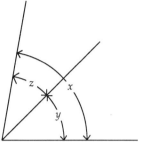 If angle x is 80° and angle y is 44°, what does angle z measure?	31 [.78]	61 [.98]	38	75

[a] Response rates for students who had and had not taken geometry were approximately the same as the response rate for all students in grade 11.

same types of misconceptions are evident. Since no units were numbered in this item and the most common incorrect response was the one in which the wrong unit was counted, it is likely that students made the same mistake as on item A.

The results of the other two items, given only to 11th-grade students, confirm that students do not do well on application problems in geometry. At least, they do not perform well on tests with a time limit.

Table 6.8
Pythagorean Relations

| | Percent Responding[a] | | | |
Item	Grade 7	Grade 11 (All)	Grade 11 (No Geometry)	Grade 11 (Geometry)

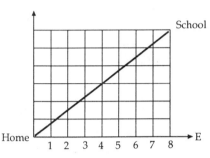

A. Jane rode 8 miles east, then 6 miles north from her home to school. If she could have gone straight from home to school, how far would she have ridden?				
6	20	19	25	15
8	32	32	43	27
10*	22	35	15	48
14	26	13	17	10
B. Find the perimeter of a rhombus drawn on grid paper.				
28		12	16	9
32		48	45	50
40*		28	26	30
48		12	13	11

[a] Response rates for item A were .53 and .98 for grades 7 and 11, respectively. Response rate for item B was .54 for grade 11.
* Indicates correct answer.

7

VARIABLES AND RELATIONS

Jane O. Swafford   Catherine A. Brown

THE use of variables and the manipulation of expressions containing variables are among the most powerful features of mathematics. Variables, in the form of boxes, are introduced in the elementary school curriculum, but it is not until the first course in algebra that these topics receive major emphasis.

The fourth assessment contained approximately fifty items intended to measure students' ability to work with mathematical variables and algebraic expressions. The items covered the following areas: *(a)* variables in equations and inequalities, *(b)* the use of variables and algebraic expressions, *(c)* variables in functions and formulas, and *(d)* coordinate systems. Only six items were administered to 3rd-grade students. Two-thirds of the items were given only to 11th-grade students, and these concerned algebraic concepts and manipulations. The rest dealt with basic concepts about variables that are included in elementary and middle school mathematics.

The data collected permitted the analysis of the 11th-grade students' performance by mathematics course background. Almost 75 percent of these students reported completing first-year algebra than any other high school mathematics course, including general mathematics. Only 40 percent of the students reported completing a second year of algebra.

As would be expected, students' performance was clearly related to course background. The average difference in performance between students who had taken at least one year of algebra and those 11th-grade students who had no algebra course was about 30 percentage points. The difference in performance between students who had taken a second course in algebra and students with at least one year of algebra averaged 9 percentage points. It should be noted that the Algebra 1 group included all students who had indicated they had taken Algebra 2. Hence, the difference between performance of students who had completed one and two courses in algebra might be more pronounced than the 9-point spread would suggest. It is apparent, however, that little of algebra is intuitive. Students must study algebra to learn algebra. It is also reasonable to expect that students who have taken additional algebra have better algebra skills than students with only one course in algebra.

HIGHLIGHTS

- Most 3rd-grade students could solve simple open sentences but had difficulty when the variable appeared first or when the use of an inverse operation was required.

- Students who had taken Algebra 1 or 2 did better than students who had not, and students who had taken Algebra 2 did better than those who had taken only Algebra 1. Algebra students, however, had difficulty with technical vocabulary, graphic representations, and all but the most routine problem situations.

- Almost all students who had either Algebra 1 or 2 could correctly solve a linear equation in one unknown but had difficulty with items involving the term *equivalent equations*.

- Algebra students were generally successful at simplifying linear algebraic expressions. About two-thirds of the Algebra 2 students and half of the Algebra 1 students could correctly translate word problems into algebraic expressions. At most one-fourth of the 11th-grade students without algebra answered these items correctly.

- Students with or without algebra demonstrated some familiarity with inequalities but had difficulty with symbolic and graphic representations of inequalities and all but the simplest, most familiar problem situations involving inequalities.

- Students in the 7th and 11th grades demonstrated some intuitive knowledge of functions but were less successful in situations that used functional notation or probed their understanding of functional concepts.

- Students at both grade levels, with and without algebra, had difficulty interpreting or manipulating formulas.

EQUATIONS AND INEQUALITIES

Equations

Simple open sentences are part of most elementary and middle school students' experience. Six items were included in the 3rd-grade assessment and two items in the 7th-grade assessment. The results suggest that most 3rd-grade students could solve simple open addition sentences that involved only basic addition facts. They had more difficulty with certain subtraction sentences and sentences that required a transformation, such as an inverse operation. The 7th-grade students were more successful on these more difficult sentences. Sample items are given in table 7.1. The format of items B and C might have affected performance. Consistent with some past research, a sentence beginning with a variable proved more difficult than one with more difficult arithmetic but a variable that appeared later. It should be noted, however, that item C appeared at the end of a test block and had a low response rate.

Table 7.1
Simple Open Sentences

	Percent Correct [Response Rate]	
Item	Grade 3	Grade 7
A. $4 + \square = 9$	94 [.94]	—[a]
B. $27 = \square + 14$	49 [.56]	90 [.95]
C. $\square - 8 = 8$	24 [.43]	60 [.62]

[a]Not administered in grade 7.

Much of beginning algebra is devoted to developing techniques for solving linear equations. The only item requiring students to solve a linear equation in one unknown that was given to 11th-grade students appears in table 7.2. Almost all the students with one or two years of algebra correctly solved this simple equations. Fewer than half the students who had not taken a first-year algebra course were successful. Although students who had studied algebra could compute the solution to an equation, they demonstrated less knowledge of the technical vocabulary of equation solving. About half the students with one year of algebra and two-thirds of those with two years of algebra knew that equivalent equations have the same solution or could identify a pair of equivalent equations. Only about one-third of the students without algebra answered these items correctly.

No computational items involving quadratic equations or systems of linear equations were included in the fourth assessment. Items concerning graphs of quadratic equations are discussed in the section on coordinate systems.

Table 7.2
Linear Equations in One Unknown

	Percent Correct [Response Rate]		
Item Description	No Algebra	Algebra 1	Algebra 2
A. Solve: $6x + 5 = 4x + 7$	43 [.94]	83 [.98]	91 [.99]
B. Identify the pair of equations which are equivalent.	30 [.90]	50 [.97]	60 [.97]

Inequalities

The assessment included several items assessing basic concepts of the order relation and inequalities that might be expected of all students, whether or not they had studied algebra. These concepts included the basic properties of inequalities, the use of appropriate notation, the graphing of inequalities, and the solution of simple problems containing inequalities. The assessment did not include items assessing students' ability to solve inequalities of the type generally covered in algebra course.

Students' performance indicated they had some familiarity with inequalities but lacked a clear understanding of them. They had difficulty with symbolic and graphic representations of inequalities and all but the simplest problem situations involving inequalities. For example, the performance of 7th-grade students on the two items given in table 7.3 illustrates the difficulty they had with symbol representations. These two items assessed students' understanding of the transitivity of order relations.

Table 7.3
Properties of Inequalities

	Percent Correct [Response Rate]	
Item Description	Grade 7	Grade 11
A. $5 < \Box$ and $\Box < 9$ implies $\Box$ could be 7.	60 [.71]	—[a]
B. $x < 9$ and $y < x$ implies $y < 9$.	38 [.83]	72 [.92]

[a] Not administered in grade 11.

Two items asking students to solve problems using the "less than" relation are presented in table 7.4. The additive situation presented in item A was less difficult for 11th-grade students than the subtractive situations presented in item B. The 7th-grade students' performance, even on the additive situation, was only slightly above chance.

Table 7.4
Solving Inequalities

	Percent Correct [Response Rate]	
Item Description	Grade 7	Grade 11
A. If x is less than 8, what must be true of $x + 6$?	22 [.63]	46 [.80]
B. If y is less than 8, what must be true of $14 - y$?	—[a]	27 [.76]

[a] Not administered in grade 7.

The 11th-grade students were also asked to solve the two multiple-choice items containing inequality symbols that are presented in table 7.5. Most of the algebra students correctly chose the largest whole number that would

Table 7.5
Symbolic Inequalities

Item Description	Percent Correct [Response Rate]		
	No Algebra	Algebra 1	Algebra 2
A. Identify the greatest whole number that satisfies $15 + N < 20$.	46 [.97]	79 [.99]	85 [.99]
B. $x - y > x + y$ implies $y < 0$.	6 [.86]	38 [.90]	50 [.91]

satisfy the inequality in item A, whereas fewer than half the students with no algebra could do so. On item B, only 50 percent of the students with some Algebra 2 chose the correct foil. Almost half of those with no algebra chose "I don't know." These items further illustrate the limited understanding that students have of inequality.

Several items assessed students' understanding of graphs of inequalities. Few 7th-grade students could correctly identify the graph of the solution of the simple inequality $3x \geq 15$. Two-thirds of the 11th-grade students could identify the correct graph. When they were asked to identify the equation or inequality that would describe the two-dimensional graph of $x \geq y$, only one-third of 11th graders responded correctly.

Although there was limited coverage of inequalities on the assessment, the results indicate that even students with one or two years of algebra did not have a very highly developed understanding of inequality. They had difficulty in reasoning with inequalities in all but the simplest and most familiar situations.

VARIABLES AND ALGEBRAIC EXPRESSIONS

Simplifying Algebraic Expressions

Students who had taken a course in algebra were generally successful at simplifying linear algebraic expressions. Several examples are presented in table 7.6. From 70 percent to 85 percent of the students with one year of algebra and 80 percent to 95 percent of the students with two years of algebra were successful on the least complex simplification items. Perform-

Table 7.6
Simplifying Algebraic Expressions

Item Description	Percent Correct [Response Rate]		
	No Algebra	Algebra 1	Algebra 2
A. $3x + 2y + 5x$	26 [.98]	86 [.98]	93 [.99]
B. $9(1 + 5x) + 13$	14 [.95]	74 [.98]	87 [.99]
C. $2 - 4(5 - x)$	12 [.93]	51 [.98]	67 [.99]

ance tended to decline as the expression became more complicated. On item C, only half of the students with Algebra 1 and two-thirds of the students with Algebra 2 got the correct answer. For this item, the response $18 + 4x$, correct except for a sign, accounted for most of the incorrect responses, which does indicate a knowledge of the distributive property. Students with no algebra had difficulty with all the items. On average, only one-fourth of the students who had not taken first-year algebra got the correct answer. The formal student of algebra appears to be essential to the development of skill in the manipulation of algebraic expressions.

Exponents and Radicals

Only two items involving exponents and radicals were included in the assessment. Most students who had taken a course in algebra could evaluate a rational expression with exponents in the numerator. Only one in five students without a course in algebra could do so.

Algebra students performed less well on an open-ended item that required them to give the square root of x^{12}. Only one in four of the Algebra 2 students and one in five of the Algebra 1 students gave the correct response. Of those who had not had an algebra course, only one in ten gave the correct answer. It should be noted, however, that the scores on open-ended items are often lower than scores on similar multiple-choice items.

Translation

A series of items asked students to translate situations into algebraic expressions or equations or to identify the situations that might fit a given algebraic equation. Algebra students were better able to translate situations into algebraic expressions and equations than students who had not had a first-year algebra course. On average, only one-fourth of the students with no algebra answered the items correctly, whereas about half of the students with Algebra 1 and two-thirds of the students with Algebra 2 did. The algebra students were most successful on an item asking them to identify a real-world situation that best fit a given algebraic equation. However, only half of those with two years of algebra chose the correct equation to describe the situation "the number of chairs (C) is twice the number of students (S)." Algebra 2 students also had difficulty in translating the following situation:

> Jim has 5 fewer marbles than Karen. If M represents the number of marbles Jim has, which expression describes the number of marbles Karen has?

Over one-third of the algebra students chose $M - 5$ as the correct answer. This response suggests that students were depending on word clues (*fewer* means *subtract*) rather than on a careful reading of the problem to guide their choice.

VARIABLES AND RELATIONS

Using Variables to Express Mathematical Concepts

Variables provide a convenient way to express general mathematical properties and relationships. Several items required students to use variables to represent basic properties of numbers. One item presented different expressions involving the multiplication or addition of a variable by 1 and 0. About 80 percent of the Algebra 2 students and 70 percent of the Algebra 1 students chose $x \cdot 0 = 0$ as the sentence that is true when any number is substituted for x. Only about 20 percent of the students with no first-year algebra chose the correct response. This performance is comparable to performance at the 7th-grade level. It appears that students' ability to represent relations with variables does depend on whether or not they have had a course in algebra.

Whereas almost 80 percent of the 11th-grade students could identify the missing numerator in a pair of equivalent fractions made up of whole numbers, they had difficulty expressing the relationship between a pair of equivalent fractions using variables. Only about one-third of the 11th graders recognized that $a/b = 5a/5b$. About 40 percent of the students with Algebra 1 and 45 percent of the students with Algebra 2 chose the correct response. Almost as many algebra students chose $a/b < 5a/5b$ as the correct response.

FUNCTIONS AND FORMULAS

A variety of items were included in the test to assess students' knowledge and understanding of functions. Students were asked to identify functions and to evaluate them both with and without function notation. Students were also asked to interpret formulas.

Functions

Students demonstrated some intuitive knowledge of functions, but they were less successful in situations that used functional notation. Parallel items were administered to assess students' ability to evaluate an expression like $a + 7$ with and without functional notation. The results are presented in table 7.7. Both 7th and 11th graders were able to evaluate the expression when functional notation was not used. When functional notation was introduced, the item proved to be more difficult.

Algebra students also demonstrated some degree of facility in computing specific values of functions. Their performance on the broad range of functional concepts, however, indicated a limited understanding of the topic. On one item, 11th graders were presented with a function machine that doubled each whole number dropped into the input slot. Students were asked to choose the best description of the output. Fewer than one-third of the algebra students could correctly identify the range. In another

Table 7.7
Evaluating Functions

Item Description	Percent Correct [Response Rate]			
	Grade 7	Grade 11 No Algebra	Grade 11 Algebra 1	Grade 11 Algebra 2
A. What is the value of $a + 7$ when $a = 5$?	77 [.90]	72 [.89]	94 [.94]	96 [.98]
B. If $f(a) = a + 7$, what is the value of $f(5)$?	—[a]	31 [.52]	65 [.77]	79 [.84]

[a] Not administered at grade 7.

item, students were presented four graphs and asked to choose the graph that did not represent a function. Two-thirds of the Algebra 1 students and half of the Algebra 2 students chose the graph of a circle. But when asked to identify the graph that showed the graph of a function and its inverse, fewer than 20 percent of the students in any category got the correct response.

Formulas

Mathematics is most often encountered in other subjects in the form of formulas that express relationships. Two items were included in the assessment to test students' ability to substitute values into a formula and to interpret a formula. On both tasks performance was low. The formula was the relationship between Fahrenheit and Celsius temperatures. When asked what the corresponding increase in degrees Fahrenheit would be for each increase of one degree Celsius, fewer than 30 percent of the algebra students chose the correct response. Performance was about the same when students were asked to substitute a value into the formula. On this task, 40 percent of the Algebra 2 students and 30 percent of the Algebra 1 students were successful.

COORDINATE SYSTEMS

The interplay between geometry and algebra strengthens students' ability to formulate and analyze problems within and outside mathematics. A number of items were included that assessed students' understanding of the representation of algebraic concepts in a Cartesian coordinate system. These items included such tasks as identifying points on a coordinate system, identifying a linear equation from its graph, and answering questions about the graphs of quadratic equations, exponential functions, and trigonometric functions.

Most algebra students were successful at giving the coordinates of a point in the third quadrant on a standard coordinate system. About 50 percent of the 7th-grade students and 60 percent of the 11th-grade students without first-year algebra could also identify the point. However, when they were asked the coordinates of a point 2 units to the right and 3 units up from a given point, only 50 percent of the algebra students and 30 percent of the students without an algebra course answered correctly.

A number of additional items on coordinate systems were given, but they appeared at the end of the blocks. The response rates on these items were too low to try to draw any conclusions about students' knowledge of the relationship between algebraic equations and their graphs.

The assessment clearly indicated that algebra is learned in algebra classes. Eleventh-grade students who had not taken either Algebra 1 or Algebra 2 consistently performed at the same level as 7th-grade students on items involving variables and functions. However, the performance of those students who had taken algebra indicated that although they had learned symbol manipulation, they were unable to use variables and relations except in the most routine problem situations.

8

NUMBER AND OPERATIONS

Vicky L. Kouba   Thomas P. Carpenter
Jane O. Swafford

SINCE numbers and operations dominate much of the mathematics curriculum, this content area received substantial emphasis in the fourth assessment. About 50 percent of the total number of items administered to 3rd-grade and 7th-grade students dealt with number and operations. For 11th-grade students, the figure was about 40 percent.

Knowledge and understanding of concepts, computational skills, routine applications, estimation, and problem solving were assessed for whole numbers, fractions, decimals, and percents. Items assessing knowledge of number and operations required simple recall, and those assessing skills involved computations with various types of numbers. Students' understanding of number and operations was examined through items that required them to translate from one form to another (e.g., write a number sentence to describe a model) or to explain steps in a procedure. Routine application items included familiar word problems and consumer problems. Estimation items involved calculations and word problems, and problem solving items were nonroutine problems not normally encountered in the mathematics curriculum.

This chapter contains sections on whole numbers, fractions, decimals, percents, other number topics, and estimation. Not all items are reported, but representative items are given to support major findings. Limitations in the number and kind of items are also discussed.

WHOLE NUMBERS

Perhaps no other topic in the school mathematics receives more attention in the curriculum or time in the classroom than whole numbers. The fourth mathematics assessment contained about ninety items that dealt with whole number concepts and properties, whole number operations, and word problems that used whole numbers. The following conclusions summarize the findings of the assessment; they are discussed in more detail in the three subsections.

NUMBER AND OPERATIONS

> **HIGHLIGHTS**
>
> • Approximately two-thirds of the 3rd-grade students appeared to have mastered place-value notions involving tens, but fewer than half were successful with tasks involving place-value notions beyond tens.
>
> • Over 80 percent of the 3rd-grade students could perform simple two-digit addition and subtraction computation with whole numbers, but fewer than half of them were successful in problems with more than two addends.
>
> • Most 7th-grade and 11th-grade students were successful on simple whole number addition, subtraction, and multiplication computations. Whole number division was not assessed.
>
> • Students at all three grade levels were more successful on one-step word problems containing whole numbers than on two-step word problems. Students generally tried to solve two-step problems by using only one step.

Whole Number Concepts and Properties

This section focuses on items that dealt with number and numeration concepts, order, properties, and number theory. Emphasis at the 3rd-grade and 7th-grade levels was on number, numeration, and order. At the 11th-grade level the emphasis was on number theory. Unlike earlier assessments, items involving counting scattered objects, writing numbers from spoken words, and identifying ordinals were not included. Further, the coverage of order, properties, and number theory was also limited.

Number and numeration. An understanding of number and our numeration system is essential to any work with whole numbers. From the results of the number and numeration items, it appears that 3rd-grade students could count by twos but had not mastered place value and grouping. Generally, 3rd-grade students' performance ranged about 65 percent on items involving grouping by tens, identifying the tens digit, and finding the number that was 10 more than a given number. When a place value higher than ten was involved, performance was below 50 percent. Generally, 7th-grade students were more successful with higher place values, but almost 30 percent were unable to give the number 100 more than a given three-digit number. A selection of items and their results appear in table 8.1.

Order. A series of items given to the 3rd-grade students dealt with determining the relative size of numbers (e.g., Which number is greatest? Which number is closest to a given number?) and describing the relation-

Table 8.1
Numeration Items

Items	Percent Correct [Response Rate]	
	Grade 3	Grade 7
A. What digit is in the tens place in the number 2059?	64 [.58]	—[a]
B. What digit is in the thousands place in the number 43,486?	45 [.87]	—[a]
C. What number is 10 more than 98?	61 [.85]	—[a]
D. What number is 100 more than 498?	43 [.58]	72 [.88]

[a] The item was not administered at this grade level.

ship between two numbers with a mathematical sentence (5 < 8). Students' performance was 50 percent or better. The most difficult ordering task for 3rd-grade students was choosing the number closest to 300 when given a number line like this:

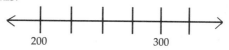

On this item almost as many students chose the number 280 as the number 310. The dual representation (verbal and pictorial) might have caused difficulty, or to these students *close* may have implied "close but no greater than."

All three age groups were given two items involving the number line. One number line, calibrated by ones, pictured *a* to the left of *b*. Only the 11th-grade students were generally successful at identifying the correct mathematical sentence for this graph. When given a number line calibrated by fours, the majority of the 7th-grade and 11th-grade students correctly identified the first point as 4. In general, 3rd-grade students and, to a lesser extent, 7th-grade students seemed to have difficulty dealing with order relationships when either variables or a nonstandard number line was involved. Both younger age groups could handle order relationships among whole numbers.

Another set of items asked for nonnumerical objects to be ordered by such characteristics as size, age, or place in line. Even the 3rd-grade students did fairly well on these items. Performance on these items seemed to be based more on general knowledge than on a knowledge of specific mathematical relations. A selection of the items involving order is presented in table 8.2.

Properties. The only items dealing with the properties of whole numbers included in the assessment focused exclusively on the properties of 0 and 1. The assessment did not include items on the commutative, associative, or distributive properties. When asked to give the value of *N* to make the

NUMBER AND OPERATIONS

Table 8.2
Order Relation

Items	Percent Correct [Response Rate]		
	Grade 3	Grade 7	Grade 11
A. Given four 4-digit numbers, which is greatest?	70 [.97]	—	—
B. Given $8 > 6$, $8 = 6$, $8 < 6$, which one is true?	60 [.96]	—	—
C. Given a number line with a to the left of b, which is true: $a = b$, $a < b$, $a > b$?	23 [.81]	36 [.76]	68 [.97]
D. Given H is taller than B and B is taller than P, identify that H is taller than P.	51 [.55]	76 [.98]	87 [.98]

sentence $15 \times N = 15$ true, slightly more than half of the 3rd-grade students correctly named 1. In the upper two grades, items requiring a higher-level application of the properties of 0 and 1 were included. These students were asked to identify which numbers would work in the expression $M \times M = M$. At the 7th-grade level, almost half identified both 1 and 0; at the 11th-grade level, almost three-fourths did. At both grade levels, one-fourth of the students identified only one of the two correct answers.

Number theory. The number-theory items fell into two categories: those that dealt with even and odd numbers and those that dealt with multiples or factors. No items dealt directly with primes.

Over 90 percent of the 7th-grade students could identify an even number. Also, 70 percent of the 7th-grade students and 80 percent of the 11th-grade students knew that the sum of two even numbers is always an even number. However, fewer than 70 percent of the 11th-grade students knew that if n is even, $n + 2$ is also even. Approximately 40 percent of the 3rd-grade students knew that if n is odd, $n + 1$ is not. One item, presented in table 8.3, asked

Table 8.3
Number Theory Item

Item	Percent Responding[a]	
	Grade 7	Grade 11
What is always true about an even number?		
It is divisible by 4.	19	10
It is one less than some odd number.*	42	60
It is always prime.	25	23
It is less than 10,000.	1	1
I don't know.	14	7

[a] The response rate was .95 for grade 7 and .98 for grade 11.
*Indicates correct response.

which of a list of statements was always true about an even number. Fewer than 50 percent of the 7th-grade students and only 60 percent of the 11th-grade students knew that an even number is one less than some odd number. What is surprising is that about one-fourth of each group thought that an even number is always prime.

About half the 3rd-grade and three-fourths of the 7th-grade students could apply number theory concepts about multiples to solve problems. Three-fourths of the 7th graders could also identify a common factor for 14 and 21. In 11th grade, two-thirds could identify the correct Venn diagram for the common factors of 24 and 30.

A more difficult item on divisibility appears in table 8.4. On this item, half of each age group identified the product of the two prime divisors as another divisor. The percent of correct responses seems low for 11th-grade students. The item, however, is complicated because 3 is also a factor of 6. Solving the item correctly requires an understanding of prime factors as well as divisibility. Without that understanding, students could have easily concluded that the unknown number is also divisible by 18. One would expect the 11th-grade students to perform better than 7th-grade students, at least well enough to recognize that "divisible by 7" makes no sense.

Table 8.4
Divisibility

Item	Percent Responding[a]	
	Grade 7	Grade 11
A certain whole number is represented by B. If B is divisible by 3, 5, and 6, then which of the following *must* be true?		
B is divisible by 7.	18	15
B is divisible by 15.*	48	55
B is divisible by 18.	23	21
B is divisible by 14.	11	9

[a] The response rate was .77 for grade 7 and .77 for grade 11.
*Indicates correct response.

WHOLE NUMBER OPERATIONS

Meaning of Operations

Students' understanding of the meaning of operations was assessed using pictorial models and number sentences. No pictorial items were administered for subtraction or division. The one pictorial item for addition depicted two jumps on a number line and asked students to write an appropriate addition sentence to match the model. The item was given only to 7th-grade students, of whom about 30 percent wrote a correct sentence. It is difficult to make a judgment on the basis of one item, but because the

NUMBER AND OPERATIONS

item was given at a grade level in which students are expected to have well-developed concepts of addition, it seems that the number-line model is not part of 7th-grade students' concept of addition.

About 60 percent of the 3rd-grade students were successful on two items involving pictorial representations of multiplication. On one item students were presented an array (3 rows of 4 objects) and asked to write a number sentence to match the model. On the other item students were presented a grouping picture and asked to identify the number sentence that matched the model. It is difficult to come to a definite conclusion on the basis of these two examples; however, it appears that in spite of the fact that 3rd-grade students have had little exposure to multiplication, the majority recognized that both the array and grouping models represent multiplication.

Students' understanding of the relationship between addition and subtraction was assessed by one number-sentence item at grades 3 and 7. Performance is summarized in table 8.5. In these grades students have some background for understanding this item in that textbooks generally include exercises involving families of facts like the following:

$$3 + 5 = 8$$
$$5 + 3 = 8$$
$$8 - 3 = 5$$
$$8 - 5 = 3$$

Table 8.5
Equivalent Number Sentences

	Percent Responding[a]	
Item	Grade 3	Grade 7
If $49 + 83 = 132$ is true, which of the following is true?		
$49 = 83 + 132$	52	18
$49 + 132 = 83$	13	7
$132 - 49 = 83$*	29	61
$83 - 132 = 49$	7	14

[a] Response rates were .63 for grade 3 and .92 for grade 7.
*Indicates correct response.

Fewer than 30 percent of the 3rd-grade students responded correctly. About 50 percent of them chose the first distracter. This may have been due partly to guessing, partly to focusing on the addition, or partly to focusing on the fact that the numbers appear in the same order as in the original sentence. The size of the numbers also may have been a confounding factor, since other parts of this assessment indicate that 3rd-grade students have difficulty with three-digit numbers. Students in grade 7 did better on the item, with about six out of ten responding correctly.

COMPUTATION WITH WHOLE NUMBERS

Fourteen items were used to assess whole number computation, most of which dealt with two-digit addition and subtraction. No computation items involving whole number division were given to 7th-grade or 11th-grade students. No multiplication items that involved regrouping or zeros were given.

Eleven of the fourteen items were assessed at all three grade levels, making it possible to examine progress across the grades. These eleven items are listed in table 8.6 and are assigned a letter (A–K) for ease of discussion. A graph of performance on the eleven items is presented in figure 8.1.

In considering the students' performance, one should keep in mind that 3rd-grade students' experience in mathematics is usually with addition and subtraction of two-digit numbers, and they have only a slight familiarity

Table 8.6
Whole Number Operations

	Percent Correct[a]		
Items	Grade 3	Grade 7	Grade 11
A. 45 +32	96	96	98
B. 57 +35	84	95	97
C. 49 56 62 +88	48	80	88
D. 49 − 36	85	94	97
E. 69 − 35	86	95	97
F. 44 − 6	65	91	94
G. 54 −37	70	94	97
H. 242 −178	50	85	91
I. 504 −306	45	84	90
J. 31 × 3	56	94	96
K. 213 × 12	7	77	84

[a] Response rates for addition and subtraction items were .92 or greater. Response rates for multiplication items were .80 or greater.

NUMBER AND OPERATIONS

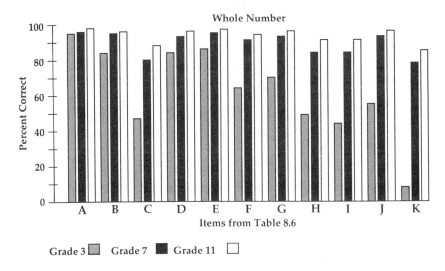

Fig. 8.1. Students' performance on eleven whole number computation items. Item letters refer to table 8.6.

with three-digit and four-digit numbers. Most 3rd-grade students have had some experience with multiplication but little experience with division. Students in grades 7 and 11, however, have had experience and practice with all four operations and should have mastered computational algorithms with whole numbers.

Addition. Three addition items were administered in all three grade levels (see items A, B, and C in fig. 8.1 and table 8.6). Students at all three grade levels did well on the two-digit addition item that did not require regrouping. Almost all 7th-grade and 11th-grade students and about 80 percent of the 3rd-grade students successfully performed two-digit addition involving regrouping. The two-digit addition problem with four addends proved the most difficult for all three grades; about 50 percent of the 3rd-grade students, about 80 percent of the 7th-grade students, and about 85 percent of the 11th-grade students responded correctly.

No other whole number addition items were given to 7th- and 11th-grade students. Students in grade 3 were given an addition item with three addends (5 + 37 + 8). About 60 percent of the 3rd-grade students responded correctly, performing a little better on this item than on the item involving four addends. Performance may have been a bit better because fewer addends were involved and the sum did not exceed 100.

Subtraction. Six whole number subtraction items were administered to all three grade levels (see items D–I in fig .8.1 and table 8.6). About 85 percent of the 3rd-grade students correctly solved the items involving two-digit subtraction with no regrouping (borrowing). Performance for 3rd-grade students fell 15–20 percentage points on two-digit subtraction items involving regrouping, and fewer than 50 percent of the 3rd-grade students could correctly solve three-digit subtraction problems.

Performance of students in grades 7 and 11 was above 90 percent on the two-digit subtraction items and about 85–90 percent on the three-digit subtraction items.

From their performance on the subtraction and addition items, it appears that many 3rd-grade students have not fully mastered regrouping in two-digit addition and subtraction. However, students continue to progress toward mastery, and most of them reach it by 7th grade.

Multiplication. Two whole number multiplication items were administered to all three grade levels (see items J and K in fig. 8.1 and table 8.6). A little over half of the 3rd-grade students could do the simpler multiplication problem, but only about one in ten could correctly multiply a three-digit number by a two-digit number. As might be expected, 7th-grade and 11th-grade students did better on these two items than the 3rd-grade students. However, because only two multiplication items were given (neither of which involved regrouping), little can be concluded about students' general performance on whole number multiplication.

Division. Only the 3rd-grade students were assessed on whole number division, and they were only given two items. About three-fourths of the students responded correctly to 12 divided by 4. However, fewer than 20 percent of them correctly solved 42 divided by 3. This level of performance may have occurred because many 3rd-grade students had not yet learned a computational algorithm for this type of problem.

Word Problems involving Whole Numbers

The assessment included one- and two-step word problems similar to those in textbooks, as well as a few nonroutine problems. Some items required students to choose an open sentence to represent the problem rather than to solve the problem. The items were given to one or two grade levels; no whole number word problems were assessed at all three grades.

One-step word problems. Ten one-step word problems were used. Seven were administered to 3rd-grade students, five to 7th-grade students, and two to 11th-grade students. The performance for 3rd graders ranged from 30 to 90 percent correct. Table 8.7 has a brief description of each item and the 3rd graders' performance. For five of the seven items given to 3rd graders, a comparable computation item was also given. Performance on those is also noted in table 8.7 for comparison. In general, 3rd graders performed better on the computation items than on word problems involving the same or similar whole number computations.

Table 8.7
One-Step Word Problems (Grade 3)

	Percent Correct [Response Rate]	
Item Description	Word Problem	Corresponding Calculation Problem
Addition (32 + 45)	88 [.95]	96 [.97]
Addition (30 + 35) with extraneous information	58 [.95]	
Subtraction (136 − 48) sentence	57 [.83]	
Subtraction (16 − 11)	66 [.78]	85 [.93]
Subtraction (100 − 94)	68 [.97]	70 [.95]
Multiplication (4 × 5) with extraneous information	32 [.69]	56 [.83]
Division (12 ÷ 3)	58 [.90]	77 [.68]

Although 3rd-grade students were more successful at solving a subtraction word problem than choosing a number sentence that accurately represents a subtraction word problem, the results may be misleading. As table 8.8 shows, the item assessing 3rd-grade students' ability to choose an appropriate number sentence required a comparison of two disjoint sets,

Table 8.8
Subtraction Word Problems (Grade 3)

Item	Percent Responding[a]
A. Steve has 48 bottle caps. Julie has 136. Which number sentence could be used to find how many more caps Julie has than Steve?	
136 − 48 = ☐ *	57
136 + ☐ = 48	5
48 − ☐ = 136	3
48 + 136 = ☐	13
I don't know.	22
B. A spice rack holds 16 bottles. There are 11 bottles already in the rack. How many more bottles can be put in the rack? (Correct answer: 5)	66
C. Robert spends 94 cents. How much change should he get back from $1.00?	
6¢*	68
16¢	5
65¢	3
41¢	2
94¢	4
I don't know.	17

[a] Response rates were .83 for item A, .78 for item B, and .97 for item C.
*Indicates correct answer.

whereas the items assessing their ability to solve subtraction word problems involved a whole set and its parts. The difference in performance may result from the difference in the items, since comparison or difference items generally are more difficult than part-whole items. Another confounding factor is that on the number sentence item, 13 percent of the students incorrectly chose the addition sentence. Addition is the first operation students learn and is often their first choice as a strategy for solving word problems. Thus, the number of students opting for this distracter is not unusual. However, this choice was not as readily available in the other two subtraction items. Item B was not a multiple-choice item, and item C did not have $1.94 as a choice. Because of these confounding factors and the small number of items administered, no conclusions can be made about 3rd-grade students' performance on word problems requiring number sentences compared to their performance when they just had to solve the problem.

The only one-step word problems given to both 3rd- and 7th-grade students were the two addition items. As shown in table 8.9, extraneous information in a word problem increased its difficulty. The most common error made by both 3rd- and 7th-grade students was adding all three numbers in the problem rather than just two of them. This error seems to predominate even when the appropriate operation is not addition. Table 8.10 shows 3rd-grade students' performance on a multiplication item containing extraneous information. More students chose the three-addend

Table 8.9
Effect of Extraneous Information on Performance

	Percent Responding[a]	
Item	Grade 3	Grade 7
A. Jake has 32 toy trucks. He buys 45 more. How many does he have in all?		
3	2	1
32	1	1
45	1	1
77*	88	95
87	4	3
I don't know.	5	1
B. At the store, a package of screws costs 30¢, a roll of tape costs 35¢, and a box of nails costs 20¢. What is the cost of a roll of tape and a package of screws?		
55¢	8	3
65¢*	58	78
50¢	9	3
85¢	26	16

[a] Response rates were .98 for grade 3 and .99 for grade 7 on item A and .95 for grade 3 and .99 for grade 7 on item B.
*Indicates correct answer.

NUMBER AND OPERATIONS

Table 8.10
Multiplication (Grade 3)

Item	Percent Responding[a]
Pam has 4 pictures. There are 3 trees and 5 cars in each picture. Which number sentence gives the total number of cars in the pictures?	
4 + 5	13
4 × 3	15
4 × 5*	32
4 + 3 + 5	41

[a] Response rate was .69.
*Indicates correct answer.

sentence as representative of the problem than chose the appropriate multiplication sentence.

The three other one-step word problems given to 7th-grade students involved division. Two of these were also given to 11th-grade students. The results are given in table 8.11.

Table 8.11
One-Step Division Word Problems

	Percent Responding[a]	
Item	Grade 7	Grade 11
A. Kate is packing crayons in boxes. Each box holds 8 crayons. She has 24 crayons. Which number sentence will help her find out how many boxes she will need?		
24 − 8	4	1
24 ÷ 8*	74	89
24 + 8	6	2
24 × 8	12	6
I don't know.	3	1
B. Eight students paid a total of $48 for tickets to the dance. Which shows how to find how much each ticket cost?		
48 × 8	20	
48 ÷ 8*	67	
48 + 8	10	
48 − 8	4	
C. Four pets are competing in a best pet contest. Eighty children vote and the pet with the most votes will win. What is the smallest number of votes that a pet could receive and still win the contest? (Answer: 21 votes)	28	58

[a] Response rates were about .95 for item A, .54 for item B, and .78 (grade 7) and .90 (grade 11) for item C.
*Indicates correct answer.

About 70 percent of the 7th-grade students and about 90 percent of the 11th-grade students can correctly identify division as the appropriate operation for a simple word problem. About 10–20 percent of the 7th-grade students inappropriately chose a multiplication sentence to represent a division problem, indicating that although they recognized the grouping nature of the situation, they could not correctly distinguish between division and multiplication.

Two-step word problems. Six two-step, whole number word problems were included in this assessment. Four were given to 3rd-grade students, four to 7th-grade students, and one to 11th-grade students. In general, students did better on one-step than on two-step word problems.

A comparison of two items shown in table 8.12 gives some insight into the kinds of errors made on relatively simple two-step word problems. About two-thirds of the 3rd-grade students correctly solved the one-step item, whereas fewer than one-third solved the two-step item. Both items were multiple choice, and approximately 20 percent of the 3rd-grade students chose "I don't know" on both items. The major difficulties 3rd-grade students had were making a computing error or doing only the first of the two steps. Students in grade 7 made similar errors.

Table 8.12
Effect of Two-Step Problem on Performance

	Percent Responding[a]	
Item	Grade 3	Grade 7
A. Robert spends 94 cents. How much change should he get back from $1.00?		
6¢*	68	
16¢	5	
65¢	3	
41¢	2	
94¢	4	
I don't know.	17	
B. Chris buys a pencil for 35 cents and a soda for 59 cents. How much change does she get back from $1.00?		
6¢*	29	77
16¢	20	13
65¢	6	2
41¢	6	1
94¢	17	5
I don't know.	22	2

[a] Response rates were .97 for grade 3 on item A and .65 for grade 3 and .98 for grade 7 on item C.
*Indicates correct answer.

NUMBER AND OPERATIONS

About half of the 3rd-grade students and about 90 percent of the 7th-grade students correctly solved simple two-step word problems involving successive small additions or subtractions to a given starting number, such as in the following problem:

There are 31 birds on the fence. Six take off and 3 more land. How many birds are on the fence then?

The use of small numbers, however, did not necessarily guarantee success for 3rd-grade students. For the item presented in table 8.13, only one-third of the 3rd-grade students correctly responded. A close examination of the item reveals that it is a far more complex problem than the size of the numbers indicates. The text of the item does not state that Meg filled three boxes and *then* had six cupcakes left. The students were expected to derive this information from the picture. The use of the picture as part of the problem is not normally done in textbooks. Textbook illustrations generally are superfluous or depict the entire action of the story, not just the initial state. Thus, the students could have assumed that the picture depicted the end result of the story, making none of the offered choices correct. That "9" was the response given most often suggests that many students either treated the picture as the end result and made the error of responding with the total number of objects rather than the number of groups or simply added the two numbers in the problem.

Table 8.13
Two-Step Nonroutine Word Problem (Grade 3)

Item	Percent Responding[a]
Meg is putting cupcakes in boxes. She puts 3 cupcakes in each box. If she has 6 more cupcakes, how many boxes of cupcakes will there be in all?	
10	15
9	43
5*	32
6	10

[a] Response rate was .85.
*Indicates correct answer.

Of the two other two-step, whole number word problems administered, one was given only to 7th-grade students and the other was given to both 7th- and 11th-grade students. The results, given in table 8.14, show that the 7th-grade students did not do particularly well. Both items had multiple-choice formats, but since the distracters did not include the sum of the numbers that appeared in the problem, if the students added the numbers given in the problem, as they did for one-step problems with extraneous information, they had no appropriate choice. Most students made errors because they did only one step. For example, in item A they subtracted 4 from 40 or took one-half of 40. In item B the most frequent error involved multiplying 3×51 or 7×51. For item B in table 8.14, students may have recognized the relationship in the item as multiplication, then chose two numbers in the problem to multiply. In any event, many students attempted to solve two-step problems by performing only one operation. These results also may support the conclusion that students do not *read* long or complicated word problems.

Table 8.14
Two-Step Word Problems

	Percent Responding[a]	
Item	Grade 7	Grade 11
A. There are 40 people in the room. Four people have no opinion and the rest are split equally into those who like chocolate and those who don't like chocolate. How many people like chocolate?		
26	21	
20	18	
18*	54	
10	7	
B. After 3 days the school reported a total of 51 absences. If the absentee rate keeps this average, how many total absences will be reported after the seventh day?		
17	7	5
51	5	2
119*	42	69
153	15	6
170	4	2
357	14	9
I don't know.	14	7

[a] Response rates were .52 for item A, .85 for grade 7 on item B, and .96 for grade 11 on item B.
* Indicates correct answer.

FRACTIONS

Approximately twenty items assessed fraction concepts and operations. Three of the items were administered to 3rd-grade students. Most of the remaining items were administered to both 7th- and 11th-grade students. The three items administered to the 3rd-grade students tested simple fraction concepts. The items administered to 7th- and 11th-grade students included computation and simple concepts. The computation items were limited to subtraction and multiplication; no problems measured students' understanding of fraction operations, and three simple word problems involved operations with fractions.

HIGHLIGHTS

- Many students appear to have learned fraction computations as procedures without developing the underlying conceptual knowledge about fractions.
- Students in all three grades had difficulty with items that did not involve routine, familiar tasks, even when the items tested basic concepts.
- Slightly more than half of the 3rd-grade students could identify what fractional part of a figure was shaded.
- Slightly more than half of the 7th-grade students and about three-fourths of the 11th-grade students could perform basic computations with fractions.
- An extremely limited knowledge of fraction procedures was demonstrated by about one-third of the 7th-grade students and one-fourth of the 11th-grade students.

Basic Fraction Concepts and Procedures

The three items administered to 3rd-grade students tested very basic concepts of fractions. About 60 percent of the students could identify the shaded figure that represented a common fraction (table 8.15). About half could write a common fraction as a numeral, but only one-fourth realized that it takes four fourths to make a whole.

For 7th- and 11th-grade students, four fraction concepts were assessed: the representation of fractions on a number line, equivalence, improper fractions, and the comparison of fractions. About 40 percent of the 7th-grade students and 70 percent of the 11th-grade students could identify the point on a number line that represented a simple fraction.

About two-thirds of the 7th-grade students could identify the larger of two fractions (e.g., 3/5 and 2/6), but their performance was much lower

Table 8.15
Identification of Fractions (Grade 3)

Item	Percent Responding[a]
Which shows 3/4 of the picture shaded?	
(vertical bars, 3 of 4 shaded)	31
(triangle, 1 of 2 shaded)	3
(square, 3 of 4 shaded)	60
(diagonal, 3 shaded)	7

[a] Response rate was .94.

when they had to identify the largest and smallest of four fractions in a simple situation (table 8.16).

About 80 percent of the 7th-grade students could change a mixed fraction to an improper fraction, but fewer than half of the 7th- or 11th-grade students recognized that 5 1/4 was the same as 5 + 1/4. This is one of several items suggesting that significant gaps exist in many students' basic knowledge of fractions. It appears that many students who are successful at routine, frequently encountered calculations have difficulty when they are asked questions about fractions that do not involve standard calculations, even when the questions involve basic fraction concepts.

Computation

Performance on the five fraction computation items is summarized in table 8.17. Although none of the calculations involved large numbers, both of the subtraction items required finding common denominators, and the three multiplication items involved both fractions and whole numbers. Consequently, the items tapped most of the different skills involved in

NUMBER AND OPERATIONS

Table 8.16
Ordering Fractions

	Percent Responding[a]	
Item	Grade 7	Grade 11
John won 5/8 of the games he played; Ted won 3/4; Jim won 9/16; and Rocky won 2/3.		
A. Which of the players had the best record?		
John	10	6
Ted*	31	59
Jim	35	9
Rocky	24	26
B. Which of the players had the worst record?		
John	7	10
Ted	9	3
Jim*	44	70
Rocky	41	17

[a] Response rates for item A were .66 and .98 for grades 7 and 11, respectively, and for item B, .64 and .98, respectively.

Note: Items A and B appeared in different test blocks and were given to different groups of students.

Table 8.17
Fraction Computation

	Percent Correct [Response Rate]	
Item	Grade 7	Grade 11
A. $3\frac{1}{2} - 3\frac{1}{3}$	53 [.62]	71 [.77]
B. $7\frac{1}{6} - 3\frac{1}{2}$	32 [.57]	45 [.73]
C. $9 \times \frac{2}{3}$	60 [.79]	76 [.97]
D. $4 \times 2\frac{1}{4}$	56 [.76]	70 [.90]
E. $4 \times \frac{1}{5}$	69 [.76]	76 [.91]

subtracting and multiplying fractions.

The subtraction item requiring regrouping was significantly more difficult than the other four items. Performance was relatively consistent over the other four items. About half of the 7th- and about 70 percent of the 11th-grade students could perform simple comparisons of fractions and computations involving subtraction and multiplication with mixed numbers.

Word Problems involving Fractions

The assessment included only three word problems involving fractions. In one routine problem, 7th-grade and 11th-grade students had to find the sum of two fractions. A picture was provided that could have been used to help find the answer. Performance on this item was about the same as the level of performance on simple calculation items; a little more than half of the 7th-grade students and three-fourths of the 11th-grade students solved the problem correctly. The other two items asked students to estimate the answer and are discussed in the section on estimation.

DECIMALS

No decimal items were administered to 3rd-grade students. A total of twenty-five items were administered to the 7th- and 11th-grade students, most of which were administered at both grade levels. Items assessing concepts focused on expressing fractions as decimals and vice versa and on comparing decimals. The computation items covered all four operations. Only one item tested students' ability to use decimals to solve problems.

HIGHLIGHTS

- There is some evidence that many students are being taught decimal computation before they have learned basic decimal concepts.
- About 60 percent of the 7th-grade students could add and multiply decimals, whereas fewer than 50 percent could subtract and divide them.
- About 75 percent of the 11th-grade students could add, subtract, and multiply decimals; about 60 percent could divide them.
- Extremely limited knowledge of basic decimal concepts and skills was demonstrated by about 40 percent of the 7th-grade students and 20 percent of the 11th grade students.

Basic Decimal Concepts and Procedures

About three-fourths of the 11th-grade students could locate a decimal on the number line, which is comparable to the number of students who could locate a common fraction on the number line.

The results for the items that assessed students' ability to write common fractions as decimals and decimals as common fractions are summarized in table 8.18. About 60 percent of the 7th-grade students and 65–70 percent of the 11th-grade students could express simple fractions as decimals. The 7th-grade students' performance was significantly lower for the item that involved an improper fraction. Students in grade 11 had less difficulty with this item. For both groups performance was lower on the item that involved changing a decimal to a common fraction, even though the fraction did not

NUMBER AND OPERATIONS   83

Table 8.18
Writing Common Fractions as Decimals and Decimals as Common Fractions

	Percent Correct [Response Rate]	
Item	Grade 7	Grade 11
Write as a decimal:		
A. $\frac{6}{100}$	60 [.90]	70 [.94]
B. $4\frac{4}{10}$	59 [.86]	65 [.91]
C. $\frac{138}{100}$	38 [.86]	62 [.91]
Write as a fraction:		
D. .037	48 [.83]	58 [.77]
E. .$\overline{66}$	7 [.83]	23 [.78]

have to be reduced. This was the only item that involved thousandths, which may account for part of the difficulty. But it would be inappropriate to conclude, on the basis of a single item, that students have greater difficulty with decimals beyond hundredths. At both grade levels students were generally unfamiliar with common repeating decimals.

The results for the items involving the comparison of decimals are summarized in table 8.19. For 7th-grade students, performance on these items was significantly below their performance on the items that involved writing decimals as fractions; for 11th-grade students, performance was actually somewhat higher.

Table 8.19
Comparing Decimals

	Percent Correct [Response Rate]	
Item	Grade 7	Grade 11
A. Which number is GREATEST? (Choices: 0.36, 0.058, 0.375, 0.4)	47 [.98]	77 [.99]
B. Which number is between .03 and .04? (Correct response: .035)	35 [.88]	73 [.96]
C. Sue used a hand calculator to divide 15 by 4. She got 3.75 for an answer. This number is between which of the following pairs of numbers? (Correct response: 3 and 3 1/2)	54 [.97]	84 [.99]

Decimal Computation

Results for the decimal-computation items are summarized in tables 8.20 and 8.21. Students in grade 7 were more successful adding and multiplying than subtracting or dividing. For the 11th-grade students, performance was more consistent over the four operations.

Table 8.20
Addition, Subtraction, and Multiplication of Decimals

	Percent Correct [Response Rate]	
Item	Grade 7	Grade 11
A. 6.002 + .02 + 100.4	59 [.91]	83 [.96]
B. 4.3 − .53	43 [.92]	65 [.96]
C. Subtract 3.56 from 14.8	48 [.94]	76 [.93]
D. 7.2 × 2.5	62 [.96]	76 [.98]
E. .2 × .4	58 [.97]	78 [.99]

Table 8.21
Decimal Division

	Percent Correct [Response Rate]	
Item	Grade 7	Grade 11
A. $3\overline{)9.06}$	52 [.94]	67 [.97]
B. $.02\overline{)8.4}$	46 [.91]	67 [.94]
C. 98.56 ÷ .032	36 [.72]	60 [.81]
D. Divide .0884 by 3.4	36 [.70]	55 [.80]

Every division item had zeros in the quotient, so the computations were relatively difficult. Given the range of subskills required in the division problems, performance was remarkably consistent over the four problems, particularly in the 11th grade. The results suggest that most students who have learned to divide decimals have learned the broad range of subskills necessary to solve different types of problems, but about a third of the 11th-grade students have learned very little about dividing decimals and cannot solve the most basic decimal items.

One approach for teaching decimal concepts and operations would be to establish basic concepts involving the meaning of decimals and then build computational skills on this understanding. If this were the pattern, one might expect that performance in the 7th grade would be highest on the items testing students' ability to compare decimals and write fractions as decimals and the greatest gains between the 7th and 11th grades would be on computation items. This was not always the case. Students generally were less successful comparing decimals than they were on all but the most difficult division items. The success rates on the different items suggest that most students at both grade levels who have mastered basic concepts of decimals can also perform the computations. It does not appear that the primary focus of instruction at the early grade levels is on developing basic decimal concepts. The success rates on the items testing basic concepts suggest that many students may be attempting to learn decimal computation before they have learned basic decimal concepts.

In the one word problem involving decimals, 11th-grade students were asked a problem similar to the following:

Mary walks .8 kilometers every day. How many kilometers does she walk in 35 days?

This problem provides a familiar context involving the multiplication of a decimal by a whole number, and students were generally successful. About 80 percent of the 11th-grade students correctly answered the question. This rate is slightly higher than performance on any decimal multiplication item, but none of the computation items were exactly comparable to the computation in this problem.

PERCENTS

A total of twenty-eight percent items were administered involving concepts, calculations, and problem solving. Most of the items covering basic concepts tested students' ability to express decimals and fractions as percents and percents as decimals and fractions. The calculation items included the three basic types of percent problems and included percents ranging from a fraction of a percent to percents greater than 100. Both simple problems that required only one-step calculations and multiple-step problems were included. The multiple-step problems generally required students to draw on their knowledge of the problem situation. For example, they needed to recognize that the total amount to be repaid on a loan includes both the interest and the principal.

HIGHLIGHTS

- Students were more successful in performing calculations involving familiar percents like 25 percent or 50 percent than they were with less commonly encountered percents.
- Students were about as successful in solving simple one-step word problems with percents as they were in performing the corresponding calculations. Multiple-step word problems were significantly more difficult.
- About one-third of the 7th- and two-thirds of the 11th-grade students could perform basic calculations involving percents.

Basic Concepts

The results for selected items in which students were asked to express percents as fractions or decimals and decimals as percents are summarized in table 8.22. The item in which students were asked to express .42 as a percent probably does not provide a very good measure of students'

Table 8.22
Meaning of Percent

Item	Percent Correct [Response Rate]	
	Grade 7	Grade 11
Writing decimals as percents:		
A. .42	68 [.98]	84 [.99]
B. .9	31 [.68]	54 [.98]
Writing percents as decimals:		
C. 12%	71 [.98]	90 [.99]
D. .9%	25 [.79]	56 [.96]
Writing percents as fractions:		
E. 25%	43 [.97]	77 [.99]

understanding of the relation between decimals and percents. Simply dropping the decimal point and writing the percent sign produces the correct answer. Performance was much poorer when students were asked to write .9 as a percent. About one-third of the 7th-grade students and one-fourth of the 11th-grade students chose 9 percent as the response for this item. Students were more successful in changing 25 percent to a fraction than they were for some translations involving decimals. Although changing a percent to a fraction includes an extra step, the results for this item suggest that students are more successful working with common familiar percents for which they know fraction equivalents.

About 70 percent of the 11th-grade students recognized that 4 percent was equivalent to 4 out of a 100, and about 70 percent of the 7th-grade students and 90 percent of the 11th-grade students recognized that if a quantity is partitioned into parts represented by percentages the sum of the percentages must total 100 percent. Overall, it appears that about one-third of the 7th-grade students and a little more than half of the 11th-grade students have learned basic percent concepts, although performance was somewhat higher on some simple tasks.

Percent Calculations

The results for the items involving calculations with percents are summarized in table 8.23. The results for the two items in which students were asked to find what percent one number is of another number suggest that students are significantly more successful working with familiar percents like 50 percent than they are with less frequently encountered percents like 7 percent. They seem to be able to solve simple problems involving familiar percentages for which they know fraction equivalents without relying on computational procedures.

Not surprisingly, students were more successful calculating a percent of a number than solving other types of percent problems. Their overall

NUMBER AND OPERATIONS

Table 8.23
Percent Calculation

Item	Percent Correct [Response Rate]	
	Grade 7	Grade 11
A. 4% of 75	32 [.94]	62 [.84]
B. 76% of 20 is greater than, less than, or equal to 20?	37 [.93]	69 [.99]
C. 30 is what percent of 60?	43 [.92]	70 [.99]
D. 9 is what percent of 225?	20 [.80]	34 [.89]
E. 12 is 15% of what number?	22 [.47]	43 [.91]

success in calculating the percent of a number was about equivalent to their success in expressing percents as decimals and decimals as percents.

Percent Problems

Results for selected word problems containing percents are summarized in table 8.24. Students were about as successful in solving simple one-step word problems as they were in performing similar calculations; however, many multiple-step word problems were much more difficult, even when the additional step required relatively simple arithmetic. Although item B requires several steps to find the total number of games played and then to subtract the number won to find the number lost, the extra step is relatively

Table 8.24
Word Problems with Percents

Item	Percent Correct [Response Rate]	
	Grade 7	Grade 11
Simple One-Step Problems		
A. Robin's baseball team lost 75% of its games last year. If they played 40 games, how many games did they lose?	38 [.81]	68 [.99]
Multiple-Step Problems		
B. A baseball team won 10 games, which was 40% of the total games played. If there were no ties, how many games did the team lose?	—	42 [.97]
C. Jesse borrowed $650 for one year to buy a motorcycle. If she paid 18% simple interest on the loan, what was the total amount she repaid?	9 [.73]	37 [.85]
D. An advertisement for a bicycle says that the price of the bicycle has been reduced by 60% of its regular price. If the sale price of the bicycle is $136, what was the regular price?	2 [.53]	5 [.62]

obvious, and performance on this item was at about the same level as performance on corresponding calculation items (see item E in table 8.23.).

Item C, however, was much more difficult than item A, even though they both required computing a percent of a number. Although the problem involving the repayment of a loan may not appear to require any great insights about high finance, it appears that many students treated the problem rather mechanically, without thinking about the fact that the repayment should include both the principal and the interest. It is possible that many students did not think about the problem as a real-world situation.

The last problem involved several steps along with a relatively difficult calculation. It also required some attention to the problem situation to recognize that the price had been reduced 60 percent, not that the sale price was 60 percent of the regular price. This problem was extremely difficult for almost all students.

It would be inappropriate to conclude from the results of items C and D that performance would be the same if students were in a real situation in which they needed to perform similar calculations, but it appears that many students may treat word problems they encounter in school and on tests without thinking about the situation they represent.

OTHER NUMBER TOPICS

Other topics assessed operations with integers, scientific notation, and square roots. Items on integers focused on whether students could identify positive and negative numbers and the operations of addition and division. About 20 percent of the 7th-grade students and 70 percent of the 11th-grade students could identify −(3) as a negative number, and about 15 percent and 75 percent, respectively, could identify −(−12) as a positive number. About 40 percent of the 7th-grade students and over 80 percent of the 11th-grade students could add a positive and a negative integer, and about 15 percent and 75 percent, respectively, could divide two integers.

About one-fourth of the 7th-grade students and half of the 11th-grade students could identify an integral approximation of a square root. Almost three fourths of the 11th-grade students could use scientific notation in simple cases.

Estimation

Efforts were made in this assessment to determine how well students could estimate answers to computations and to word problems. The difficulty with estimation items in a paper-and-pencil multiple-choice assessment is that there is no way to tell what strategies the students used. Rather than arriving at an estimation on the basis of mental computation

NUMBER AND OPERATIONS   89

using some rounding or "sensing" the relative size of the numbers in an item, students may have used the less efficient means of doing the exact paper-and-pencil calculations and then choosing the number that was closest to their answer. However, from a paper-and-pencil assessment of estimation, one can gain a measure of students' rounding skills and a sense of whether or not they look at the reasonableness of an answer.

Eighteen items were administered, including one on rounding and one on estimating the number of objects in a picture. Six items were given to 3rd-grade students, twelve to 7th-grade students, and eleven to 11th-grade students. The computation and application items involved mostly whole numbers or money, although a few involved fractions, percents, or square roots.

HIGHLIGHTS

• Students' errors on most whole number estimation application items appeared to be more a result of misinterpreting the items than of making errors in estimation.

• There was evidence that some students lacked an understanding of the relative size of numbers greater than 100.

• In general, students did poorly on the items that involved estimating using factors, percents, or square roots.

Rounding and Estimating Numerosity

Only one item given at the 11th-grade level required students to round. About 90 percent of the 11th-grade students could round a five-digit number to the nearest thousand.

Another set of items required students to round the terms in computation exercises. Performance on these items, which were given only to 3rd-grade students, is summarized in table 8.25. It appears that about 60 percent of the 3rd-grade students have mastered the rounding of two-digit numbers, whereas only 30 percent have done so with three-digit numbers. These results are in keeping with other findings in this assessment showing 3rd-grade students less adept with three-digit numbers than two-digit numbers. Results on the items indicate that about one-third of the 3rd-grade students probably have a flawed but consistent rounding strategy of always rounding down (or truncating or keying on the first digit). As suggested in the discussion of students' performance on the order items, students may interpret "close" to mean "close but no greater than."

The only item assessing the ability to estimate quantity was given to 3rd-grade students. They were asked to estimate the number of birds in a

Table 8.25
Rounding Items (Grade 3)

Item	Percent Responding[a]
A. 72 − 49 is closest to:	
70 − 40	34
70 − 50*	55
80 − 40	5
80 − 50	7
B. 591 + 308 is closest to:	
500 + 300	30
500 + 400	24
600 + 300*	34
600 + 400	12
C. 47 +54 Which is about the same answer?	
50* +50	62
50 +60	11
40 +50	28

[a]Response rates were .82 for item A, .84 for item B, and .91 for item C.
*Indicates correct answer.

picture. Approximately half chose the correct response of "between 100 and 1000." However, about one-fourth chose "I don't know," indicating a lack of feel for comparative sizes of quantities beyond 100 or a lack of estimation strategies.

Computational and Application Estimation

Whole numbers and money. Eight of the estimation items involved whole number computations in numeric form or in word problems.

Two computational estimation items were given, one to only 7th-grade students and one to both 7th-grade and 11th-grade students. Results are presented in table 8.26. Only about half of the 7th-grade students responded correctly to these items. Students in grade 11 performed about 25 percentage points higher than students in grade 7. The 50 percent performance level on item A, which is in essence a two-step problem, is similar to 7th-grade students' performance on other two-step items in this assessment. Also, the response of "25" as an addend, which about 20 percent of the 7th-grade students chose, is not unreasonable. One can imagine students looking at the first two numbers, 41 and 62, and thinking, "a little more than

NUMBER AND OPERATIONS

Table 8.26
Whole Number Computation Estimation

	Percent Responding[a]	
Item	Grade 7	Grade 11
A. 41 + 62 + 47 + 56 + ☐ = 250		
The number in the box is closest to:		
25	19	
50*	54	
75	8	
100	7	
I don't know.	12	
B. What is the estimated total weight of 185 lbs., 107 lbs., 216 lbs., 204 lbs., 169 lbs., and 168 lbs.?		
800	7	3
1000*	54	78
1300	18	12
2000	10	3
12,000	8	2
I don't know.	4	2

[a] Response rates were .85 for item A, .80 for grade 7 on item B, and .98 for grade 11 on item B.
*Indicates correct answer.

100." They could do the same with 47 and 56 and conclude, "I have a little more than 200, so the missing number must be 50 or less." This would make "25" a reasonable answer.

On item B in table 8.26, the performance of 7th-grade students is lower than might be expected, given that the item is a one-step problem that contains numbers students should know. As in the previous item, one of the other choices, 1300, is not unreasonable, given that an estimated total is 1100 pounds. What is disturbing is that nearly 10 percent of the 7th-grade students chose a response that was off by a factor of 10. Although this can be attributed to carelessness, it may also indicate that these students do not have a sense of the relative size of larger numbers. If so, they have not progressed much beyond the "number sense" level of 3rd-grade students.

The results on the six whole number and money-application estimation items are presented in table 8.27. Performance increased with grade level; however, it is impossible to tell whether this increase was the result of better estimation skills or better computational skills.

On item A, about 20 percent of the 3rd-grade students chose $1.50 as an estimate and about 10 percent chose $1.00. Either of these choices could have been based more on an error in understanding the item (simply adding the three amounts pictured) than on an error in estimation. Only about 10 percent of the 7th-grade students chose either of these numbers as estimates.

Table 8.27
Whole Number or Money-Application Estimation

	Percent Correct [Response Rate]		
Item	Grade 3	Grade 7	Grade 11
[Stamps 3 for 25¢] [Rings 1 for 30¢] [Pins 4 for 50¢]			
A. Pete bought 6 stamps, 3 rings, and 8 pins. About how much did Pete spend?	59 [.88]	85 [.97]	
B. A toy car costs $4.85. What is the smallest bill that is enough to pay for 4 cars?	38 [.95]	64 [.95]	73 [.99]
C. Stickers are 14 cents apiece. About how many can you buy for $1.00?		79 [.90]	92 [.99]
D. If a bag weighs 735 grams, about how many will a box of 4 bags weigh?		56 [.77]	80 [.99]
E. If you drive 240 miles per day and you drive every day of the week, how far will you drive in a week?		64 [.95]	80 [.99]
F. If your family eats 4 dozen donuts per month, about how many donuts will your family eat in a year?		37 [.62]	65 [.89]

On item B, about 40 percent of the 3rd-grade students, 30 percent of the 7th-grade students, and 20 percent of the 11th-grade students chose a five-dollar bill as being appropriate. Again, the error was probably not in estimating but rather the result of not carefully reading or understanding the item.

Performance on item F was about 20 percentage points lower than on items D and E, probably because item F is a two-step problem that requires the students to know how many months are in a year and how many objects are in a dozen. About one-fourth of the 7th-grade students chose 50 as an estimate. This could be the result of chance, of choosing to do only one step, or of misreading the problem. About 15 percent of the 11th-grade students also chose 50 as an estimate.

The only "out of range" choice offered on any of the items that attracted a large number of students responses was on item E. About 10 percent of the 7th-grade students thought that 10 000 miles a week was a reasonable estimate for that item, indicating either a lack of consideration for the reasonableness of the answer for the stated conditions or an inattentiveness to the number of zeros in the choices.

NUMBER AND OPERATIONS

In summary, it appeared that students' errors on whole number application estimation items were more a result of misinterpreting the items than in making errors in estimating. Thus, the items did not clearly assess estimation skills. There is evidence, based on the number of students choosing "out of range" or unreasonable estimates, that some students may not consider the reasonableness of their answers or may not understand the relative size of numbers beyond 100.

Fractions, percents, and square roots. In general, students did poorly on the four items that involved estimating using fractions, percents, or square roots.

Two items involved estimation with fractions. Both items required rounding mixed numbers to whole numbers, and neither required a very precise estimate. For one item students were asked to estimate the sum of 3 4/7 + 8 2/8. The choices were 10, 12, 14, and 24. The item required students to have some idea of the magnitude of each of the mixed numbers, but students needed only to recognize that the fractional parts would make the sum greater than 11 but less than 12. They did not need any idea of the magnitude of the sum of the fractional parts. About two-thirds of the 11th-grade students correctly estimated the sum.

The other item required students to estimate the cost of 8 1/2 grams of metal that sold for 57 cents a gram. The choices were $6, $9, $50, $60, and $90. In terms of estimation applied to fraction concepts, the item required that the students recognize that 8 1/2 is between 8 and 9 and that 57 cents is close to half a dollar. About one-third of the 7th-grade students and two-thirds of the 11th-grade students correctly estimated the cost.

About 40 percent of the 7th-grade students and 50 percent of the 11th-grade students responded correctly on the one estimation item involving percents. This performance is similar to their performance on other word problems involving percent, which provides further evidence that the students computed and then rounded rather than estimated.

Finally, about 20 percent of the 7th-grade students and 50 percent of the 11th-grade students could correctly estimate the square root of a number a little greater than 400. These levels of performance were the same as on the other square root problem in the assessment, which asked students to identify the two integers that the $\sqrt{18}$ would lie between.

9

CALCULATORS

Vicky L. Kouba   Jane O. Swafford

THE 1985–86 mathematics assessment was the third one that included items concerning calculators. The availability and use of calculators were assessed as well as students' performance with and without calculators. There is no way to tell if the students used the calculators or not, since no assessment was made of their choice of method (calculator, paper and pencil, or mental computation).

To assess students' ability to use calculators, students with "calculator" mathematics blocks were allowed to use calculators while responding to that block of their three-block booklet. (See the Introduction [chapter 1] for a description of testing blocks.) Prior to the start of that block, students were given a Sharp EL-240S calculator, which had the four basic operation keys as well as square-root and percent keys.

Instructions on using the calculator were printed in the assessment booklet: how to turn it on, enter a whole number, clear the display, enter "one dollar and twenty-five cents," and calculate "5 plus 11 equals" and "42 minus 12 equals." Students were then instructed to use the calculator to solve items 4–14 in the block.

After they had completed the items on which a calculator could be used, students were instructed (in the booklet) to lay aside the calculator, since they would have no further need for it. Many of the same items also appeared in other mathematics blocks on which students were not allowed to use calculators. Thus, on thirty-four items, performance with and without calculators may be compared.

HIGHLIGHTS

• In general, students with calculators performed better than students without calculators on computational items.

• Students with calculators were more successful on items in which the numbers and operations appeared in the order that they were to be entered into the calculator than on items requiring reordering. This was especially true for division items.

CALCULATORS

- About 90 percent of the students reported that calculators were available at home, whereas only 20 percent reported that calculators were available in mathematics classes at school.
- Students in grade 7 reported using calculators less frequently in school than did students in grade 3 or 11.
- The main use of calculators by the 7th-grade students was for checking answers, whereas students in the 11th grade had more varied uses.

Performance on Calculator Items

Six of the thirty-four items were given at all three grade levels. All six items were computational items, either addition or subtraction of whole numbers.

Whole number computation. Overall, students with access to the calculator were more successful than those without calculators. As is evident in figure 9.1, 3rd-grade students without the calculator did slightly better on the item

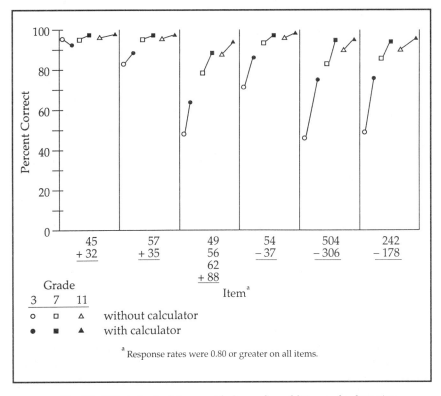

Fig. 9.1. Effect of calculators on whole number addition and subtraction

requiring addition of two-digit numbers with no regrouping. The use of the calculator had the greatest positive effect for 3rd-grade students on the items for which they probably had not mastered paper-and-pencil algorithms, such as addition with four two-digit addends and subtraction of three-digit numbers with regrouping. The performance on these items was 10 percent to 20 percent better than on corresponding items when the calculator was not used.

Four other whole number computation items were administered; two on multiplication were given to 3rd-and 11th-grade students and two on division were given only to 3rd-grade students (see fig. 9.2).

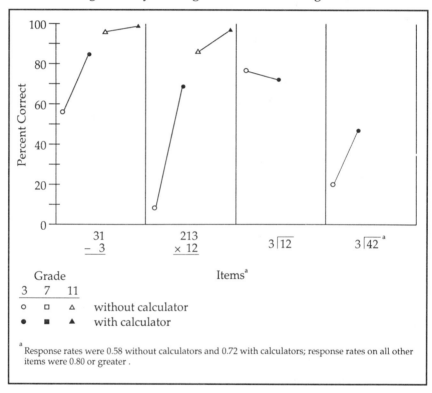

Fig. 9.2. Effect of calculators on whole number multiplication and division

About two-thirds of the 3rd-grade students with a calculator correctly computed the answer to the item in which they had to multiply a three-digit number by a two-digit number, whereas fewer than one-tenth of the 3rd-grade students without a calculator correctly answered the item. This was the greatest difference in performance between any of the pairs of calculator items, a difference probably due to 3rd-grade students' lack of experience multiplying numbers of this size with paper and pencil.

CALCULATORS

Although the difference was small, 3rd-grade students without calculators were more successful than 3rd-grade students with calculators on the division fact. On the other whole number division item given to 3rd-grade students, the difference in performance with and without calculators was much larger, about 30 points in favor of students *with* calculators. It is probable that many students chose not to use the calculator to find the simple division fact. It is also likely that those students who could not recall the division fact may have lacked a sound understanding of the division symbol and, thus, did not key in the numbers and operation in the proper order.

The performance of 11th-grade students on multiplication items was high, with little difference on the easier item (31 × 3) and a difference of about ten percentage points in favor of students who used calculators on the more difficult item (213 × 13).

Decimal computation. Eleven computation items involving decimals were administered to 7th-grade and 11th-grade students. As shown in figure 9.3, the direction and magnitude of the differences in performance with and

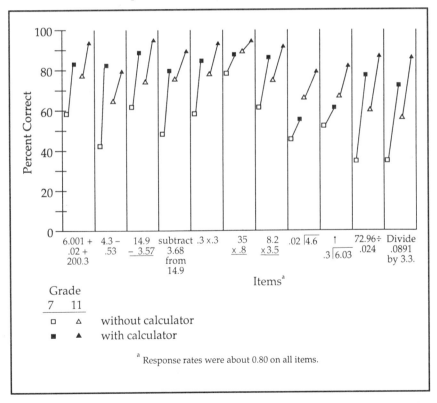

Fig. 9.3. Effect of calculators on decimal computation

without calculators were consistent across all the items. The calculator aided both age groups in accurately completing items using any of the four operations. Except for the division items that used the standard division symbol (÷) use of the calculator raised 7th-grade students' performance to levels that were equal to or beyond those of 11th-grade students without calculators and decreased the amount of difference in performance between 7th-grade and 11th-grade students. Although some of the differences might be attributable to a ceiling effect for 11th-grade students, the use of calculators still had a dramatic effect on performance. The calculator may not have been as beneficial on the items containing the division symbol (÷) because the syntax of such items does not aid students in deciding which number to enter first. For the vertical or horizontal forms of the items, numbers and symbols can be entered in the same order as they appear. Although two items appeared in verbal form, for example, "divide A by B," the phrasing suggests an order for entry of the numbers and operation into the calculator. Likewise, for the percent items and word problems discussed in the following sections, students' performance greatly increased when the syntax of the problem provided a readily identifiable (and correct) algorithm for keying in numbers and operations on the calculator.

Percent. Five items on percent were given to both 7th-grade and 11th-grade students (see fig. 9.4). Students with calculators performed better on four of the five items. The difference in performance by students with and without calculators was not as great on the percent items as on the decimal items. This may have been because the percent items could not easily be solved by pulling numbers and operations from the item and keying them into the calculator.

Performance on the item "30 is what percent of 60?" was essentially the same for both groups at each grade level, as shown in table 9.1.

Table 9.1
Simple Percent Item

	Percent Responding[a]			
	Grade 7		Grade 11	
Item	Without Calculator	With Calculator	Without Calculator	With Calculator
30 is ? % of 60?				
.5%	12	12	14	14
40%	30	30	10	9
50%*	43	38	70	67
200%	5	6	3	5
I don't know.	10	13	4	5

[a] Response rates for both grades, with and without calculators, were .92 or greater.
* Indicates correct answer.

CALCULATORS

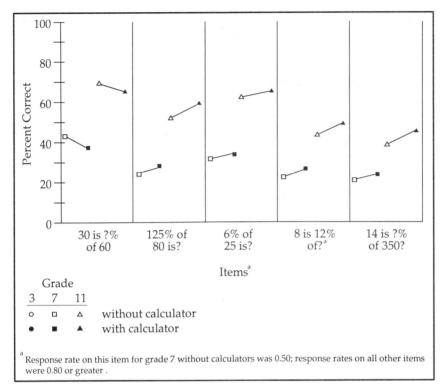

Fig. 9.4. Effect of calculators on percent computation

Because it is possible to recognize that 30 is half, or 50 percent, of 60 without doing any calculation, students may have chosen not to use the calculator (a good decision). Or they may have decided to ignore their intuitive understanding of the item and enter numbers in the calculator using an incorrect algorithm. The responses to the distracters in the multiple choices shed no light on what did occur. Again, this area needs further research: Is there a limited set of problems or an identifiable time in the learning of a specific content area when calculators are not an asset? If so, students' training with calculators should include instruction in how to make good decisions about when not to use the calculator.

Word problems. Seven word problems were administered either to just one or two of the grade levels. The students' performance on these items is reported in table 9.2 and in figure 9.5. Overall, students with calculators did somewhat better than students without calculators. The differences in performance between those with and without calculators averaged about five percentage points.

The item on which both 7th- and 11th-grade students with calculators performed 10 percent better than students without calculators was a one-

Table 9.2
Word Problems (Without and With Calculators)

	Percent Correct [Response Rate]					
	Grade 3		Grade 7		Grade 11	
Item	Without Calculator	With Calculator	Without Calculator	With Calculator	Without Calculator	With Calculator
A. Carla buys a pen for 45¢ and an eraser for 27¢. How much change does she get back from $1.00?	30 [.65]	31 [.68]	77 [.98]	81 [.97]		
B. Toy cars cost 27¢. How many can Terry buy for $2.00?			58 [.85]	54 [.85]		
C. Madge buys a new sofa for $515.00 and makes a $275.00 down payment. If she pays $21.80 a month for 12 months, how much is the total finance charge?			21 [.77]	28 [.80]		
D. Joan runs .8 of a mile every day. How many miles will Joan run in 35 days?			64 [.94]	75 [.93]	81 [.97]	91 [.97]
E. Karen's team won 75% of its games. If they played 40 games how many did they win?			38 [.81]	38 [.91]	68 [.99]	71 [.97]
F. Carl borrowed $650.00 for one year and paid 14% simple interest on the loan. What was the total amount repaid?			9 [.73]	13 [.69]	36 [.85]	48 [.91]
G. An advertisement for a sale indicated all merchandise has been reduced by 30% of its regular price. If the sale price of a stereo was $108.00, what was the regular price before the sale?			1 [.53]	1 [.66]	5 [.62]	7 [.82]

CALCULATORS

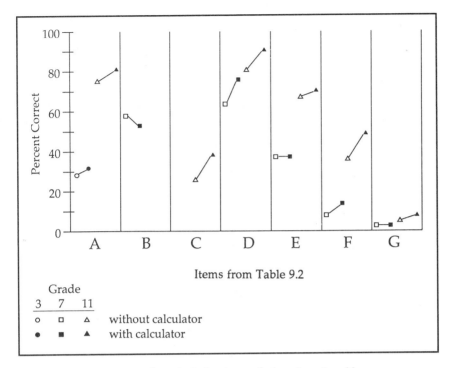

Fig. 9.5. Effect of calculators on solution of word problems

step word problem requiring multiplication of a decimal by a whole number (item D in table 9.2). The order of words in the problem—"Joan runs .8 of a mile every day. How many miles will Joan run in 35 days?"— is such that a student could enter .8, interpret *of* to mean multiplication, enter the times sign, enter the 35, press the equal sign, and get the correct answer.

This particular item is interesting for another reason: the numbers and operation in it match those used in one of the decimal computation items (see fig. 9.3). The students' performance on both these items can be analyzed by looking at the differences across both items and across the availability of the calculator, as shown in table 9.3.

Achievement differences favored the use of the calculator on both numerical and word problems. However, the differences between calculator and no-calculator users, although small, were greater on the word problem. As one can see by the differences at the bottom of each chart, the gap between performance on the word problem and the numerical item was lessened slightly by the use of calculators. The results suggest that calculators can bring performance on word problems closer to performance on numerical items; however, it is difficult to judge on the basis of one example and such small differences. This is an area for further research.

Table 9.3
Effect of Calculators on Computation versus Word Problem (35 × .8)

	Grade 7 Percent Correct		Difference
	Without Calculator	With Calculator	
Computation	80	88	+ 8
Word problem	64	75	+11
Difference	−16	−13	
	Grade 11 Percent Correct		Difference
	Without Calculator	With Calculator	
Computation	89	95	+ 6
Word problem	81	91	+10
Difference	− 8	− 4	

Availability and Use of Calculators

Calculators are quite common at home but not so readily available in the mathematics classroom, as table 9.4 shows. Students have made use of the calculator, although more 3rd-grade students than 7th-grade students reported having done so.

Students were also assessed on their use of calculators in specific school subjects and outside of school. Figure 9.6 is a graph of the percent of

Table 9.4
Availability and Use of Calculators

	Percent Responding[a]		
	Grade 3	Grade 7	Grade 11
A. "Yes" to: Do you or your family own a calculator?	82	94	97
B. "Yes" to: Is there a school calculator available for use in mathematics class?	15	21	26
C. "Never" to: How often have you used a calculator?	13	29	13

[a] Response rates were .95 or greater.

students who reported never using a calculator in mathematics, science, or any other subject or outside of school. Figure 9.7 is a graph of the percent of students reporting daily use of the calculator by subject areas.

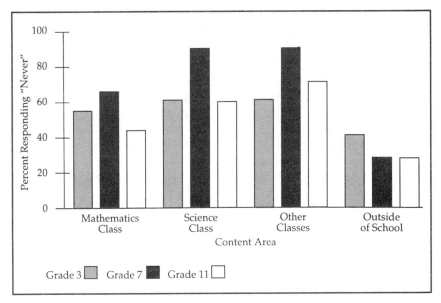

Fig. 9.6. Students never using calculators

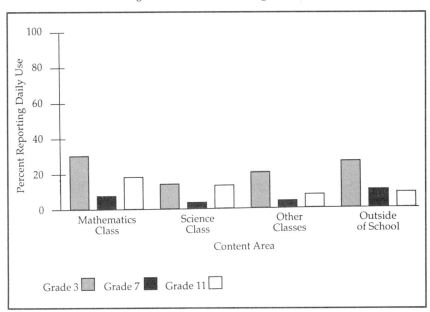

Fig. 9.7. Students using calculators daily

Although a greater percentage of 11th-grade students have used the calculator in mathematics and science and outside of school, 3rd-grade students reported more consistent use on a daily basis. They also reported that they made more use of the calculator in other subject areas than did students in grades 7 or 11. These percentages suggest that 3rd-grade students may not have understood the questions or may not have had realistic perceptions. However, the 3rd-grade students' responses may have been accurate, whereas the 7th-grade students' report of low use of calculators may simply reflect teachers' and students' perceptions that calculators are not appropriate for much of what is taught in 7th-grade mathematics. Students in grade 7 made considerably less use of the calculator in school than students in grades 3 or 11.

Table 9.5 is a summary of reported use of the calculator by 7th- and 11th-grade students. In the mathematics classroom, the uses of calculators are varied, but it is more likely to be used for checking answers.

Table 9.5
Use of Calculators in Mathematics Class

Use	Percent Responding[a]	
	Grade 7	Grade 11
Homework	30	48
Checking answers	52	52
Routine calculations	16	35
Solving problems	34	48
Taking tests	5	30
Something else	43	41

[a] Response rates were .97 or greater.

Students were also surveyed on whether they thought solving problems was more fun with a calculator. Only about 40 percent of the 3rd-grade students, 30 percent of the 7th-grade students, and 35 percent of the 11th-grade students agreed that solving problems was more fun when using a calculator.

Trends in Calculator Performance

The analysis of trends across the three NAEP mathematics assessments that included calculator items was based on performance by 9-year-olds, 13-year-olds, and 17-year-olds. Despite the evidence of success using calculators in the 1986 assessment, both the 9-year-olds and the 13-year-olds showed significant declines across time in their performance with calculators, as table 9.6 shows. Results for the 9-year-olds, based on eight calculator items that mostly involved whole number computation, showed that the drop occurred between the 1982 and 1986 assessments. For the 13-year-olds, the drop occurred between the 1978 and 1982 assessments and was based on eight decimal items. The 17-year-olds' performance initially

Table 9.6
Trends in Success in Use of Calculators

	Mean Percentages		
	1978	1982	1986
Age 9 (8 items)	78*	79*	75*
Age 13 (8 items)	59*	56	55
Age 17 (11 items)	66	63	65

*Statistically significant difference from 1986 at .05 level.

declined and then recovered. The analysis was based on 11 decimal items or word problems.

The fact that students' performance on assessment items requiring calculators has decreased over time and the reported low use of calculators in classrooms suggest that this tool is being neglected in mathematics instruction. In this age of technology, it is important that our students learn how to use the appropriate use of tools, such as calculators. School mathematics teachers have a unique opportunity to instruct students in the use of calculators, helping students recognize when and how a calculator is efficient and explore a broader range of mathematics.

10
ATTITUDES

Jane O. Swafford   Catherine A. Brown

DEVELOPING a positive attitude in students toward mathematics is as much a goal of school mathematics instruction as are achievement-related goals. Not only do we want students to be able to use mathematics, but we also want them to enjoy it, appreciate its usefulness, and understand its nature. To obtain information on students' attitudes toward mathematics, the fourth NAEP Mathematics Assessment again included a number of items to assess students' attitudes toward the subject and their perception of various aspects of mathematics learning.

Four categories of attitude items were included in the assessment: (1) *mathematics in school*, (2) *mathematics and oneself*, (3) *mathematics and society*, and (4) *mathematics as a discipline*. Items in all four categories were given to the 7th- and 11th-grade students. Only eleven items, mostly falling in the *mathematics and oneself* and *mathematics and society* categories, were given to 3rd-grade students. No attempt was made to construct attitude scales for the four categories for which a single scale score could be reported. Rather, each category contained related items that reflect the domain of interest. Therefore, the results are reported by item within the category rather than as a scale score for the category.

HIGHLIGHTS

• Among the 7th-grade students, mathematics ranked as the most important, best liked, and one of the easier academic subjects.

• Among 11th-grade students, mathematics also was one of the two academic subjects ranked highest in importance and was as well liked as the others, but it was viewed as the hardest of the four subjects they were asked to rank.

• A large majority of students at all three grade levels indicated that they wanted to succeed in their study of mathematics and were willing to work hard to do so.

- Students strongly perceived mathematics as having practical uses in the everyday world. A majority of the students believed a knowledge of mathematics is important for getting a good job, but fewer than half expected to work in careers that require mathematics.
- A large majority of the 7th- and 11th-grade students felt that mathematics is rule based but also that knowledge of process is as important as correct answers. Both groups of students lacked an understanding of what mathematicians do and of the fact that mathematics is a dynamic, cohesive discipline.

MATHEMATICS IN SCHOOL

Items in this category attempted to determine how students felt about mathematics in comparison to other subjects they encountered in school. The items were administered to the 7th- and 11th-grade students. The school subjects included in the comparison were science, social studies, mathematics, English, and physical education. Students were asked to indicate how much they liked or disliked each subject, how easy or how hard the subject was for them, and how important or unimportant they thought the subject was. A five-choice rating scale was used for each item. The results are reported in table 10.1.

Mathematics and science ranked as the best-liked subjects among the 7th-grade students. Surprisingly, physical education was liked by slightly

Table 10.1
Mathematics versus Other School Subjects

	Percent Responding[a]					
Subject	Like	Dislike	Easy	Hard	Important	Not Important
Grade 7						
Science	66	25	44	30	70	16
Social Science	56	37	46	32	67	20
Mathematics	67	26	51	31	91	4
English	59	30	57	25	84	8
Physical Education	63	12	80	5	45	39
Grade 11						
Science	64	27	39	38	68	20
Social Science	62	27	54	26	69	24
Mathematics	63	29	40	44	87	7
English	67	23	52	29	87	6
Physical Education	73	17	85	4	40	45

[a] The response rate for all items was .98 or higher.

fewer students than were mathematics and science. At the 11th-grade level, the four academic subjects were liked equally, with about two-thirds indicating that they liked each subject and one-fourth indicating that they disliked the subject. On the easy-hard dimension, more than half of the 7th-grade students rated English and mathematics as easy. At the 11th-grade level, of the academic subjects only English and social science were rated as easy by over 50 percent of the students. In fact, only 40 percent of the older group gave an "easy" rating to either mathematics or science. Although the same percentage of the group rated mathematics and science as easy, a larger percentage of the 11th-grade students rated mathematics as hard. It would appear that students' perceptions about the difficulty of mathematics change rather dramatically between 7th and 11th grade.

On the importance dimension, almost all the 7th-grade students rated mathematics as important or very important. No other subject enjoyed such a high importance rating. At the 11th-grade level, both mathematics and English were rated as important or very important by about 90 percent of the students and unimportant by only about 5 percent. One-fifth rated science as unimportant, and one-fourth rated social science low on this dimension.

Overall, mathematics was ranked as the most important, best liked, and one of the easier academic subjects by the 7th-grade students. Among the 11th-grade students, mathematics ranked as one of the two most important subjects and was as well liked as the others, but it was viewed as the hardest of the four academic subjects. This ranking is perhaps not surprising since 11th-grade mathematics is typically more difficult than 7th-grade mathematics. The curriculum moves away from arithmetic to more abstract mathematics. Although high school mathematics is admittedly more difficult, it was still viewed as important by almost 90 percent of the students and essentially was as well liked at this level as at the 7th-grade level.

MATHEMATICS AND ONESELF

Items in this category were designed to determine students' perceptions of themselves as learners of mathematics. Students in the 7th and 11th grades responded on a five-point Likert-type scales whose endpoints were "strongly agree" and "strongly disagree." (For ease in reporting, "strongly agree" and "agree" as well as "strongly disagree" and "disagree" ratings have been combined.) A smaller set of items with a three-point scale was used for 3rd-grade students. For comparison, the same set of items was also given to the 11th-grade students but not to the 7th-grade students. On the three-point scale, the response choices were "true about me," "sometimes true about me," and "not true about me." The 7th- and 11th-grade results are reported in table 10.2 and the 3rd- and 11th-grade results in table 10.3.

Table 10.2
Mathematics and Oneself (Grades 7 and 11)

		Percent Responding[a]	
Statement	Grade	Agree	Disagree
A. I really want to do well in mathematics.	7	93	3
	11	84	5
B. My parents really want me to do well in mathematics.	7	93	2
	11	85	4
C. I am willing to work hard to do well in mathematics.	7	85	5
	11	88	3
D. A good grade in mathematics is important to me.	7	89	4
	11	80	7
E. I enjoy mathematics.	7	55	25
	11	50	30
F. I like to be challenged when I am given a difficult mathematics problem.	7	55	25
	11	50	29
G. I feel good when I solve a mathematics problem by myself.	7	84	6
	11	88	3
H. I am taking mathematics only because I have to.	7	34	53
	11	26	64
I. I would like to take more mathematics.	7	44	28
	11	40	33
J. I am good at mathematics.	7	60	17
	11	53	26
K. I usually understand what we are talking about in mathematics.	7	79	9
	11	68	18
	Grade	Yes	No
L. Do you feel that you are as good in mathematics as the others in your class?	7	70	30
	11	66	24

[a] Response rates for all items were above .95.

Table 10.3
Mathematics and Oneself (Grades 3 and 11)

		Percent Responding[a]		
Statement	Grade	True	Sometimes True	Not True
A. I usually understand what we are talking about in mathematics.	3	55	36	11
	11	43	50	6
B. I am good at working with numbers.	3	65	29	5
	11	39	46	15
C. Doing mathematics makes me nervous.	3	18	27	55
	11	14	35	51
D. Mathematics is boring to me.	3	21	22	56
	11	19	48	33
E. I'm willing to work hard to do well in mathematics.	3	82	13	5
	11	51	42	8
F. I like mathematics.	3	60	22	18

[a] Response rates for all items were above .85 for grade 3 and above .95 for grade 11.

Results for grades 7 and 11. In both grades, students expressed the most positive attitudes about their desire to be successful in their study of mathematics. Over 85 percent of the 7th-grade students and over 80 percent of the 11th-grade students said that they wanted to do well in mathematics, that their parents wanted them to do well, that a good grade in mathematics was important to them, and that they were willing to work hard to achieve this goal. This strong expression of desire and motivation to do well in mathematics is interesting in view of students' less-than-impressive achievement in the subject. Is it the system or the students who are not achieving? Or do these data merely reflect that students were responding to the attitude items in the way they thought they should answer rather than the way they really felt?

Only 55 percent of the 7th-grade students and 50 percent of the 11th-grade students said they enjoyed mathematics. On the like-dislike dimension of the *mathematics in school* items reported earlier (see table 10.1), approximately two-thirds of the students at both grade levels indicated that they liked mathematics. Students may have viewed these pairs of items as very different questions or their responses may simply have been inconsistent. In any event, whereas a bare majority expressed enjoyment for mathematics, about 85 percent expressed satisfaction with solving a mathematics problem by themselves. Also, only one-fourth of the 11th-grade students said they were taking mathematics because they had to and 40 percent indicated that they planned to take more mathematics. Seventh-grade students are generally required to take mathematics, but over half indicated that this was not their only reason for taking mathematics. However, fewer than half indicated an interest in taking more mathematics.

A majority of the 7th- and 11th-grade students thought that they were good at mathematics. In fact, 70 percent of the younger students and two-thirds of the older students thought that they were as good in mathematics as the other students in their class. Also, 80 percent and 70 percent of the 7th- and 11th-grade students, respectively, claimed that they usually understood what was going on in their mathematics class. These results seem at odds with students' perception of the difficulty of mathematics as reported earlier, particularly at the 11th-grade level. Perhaps usually understanding a subject but finding it difficult are not incompatible views. But when most students think they are average or above average, the amount of effort expended to learn an admittedly difficult subject may not be adequate to produce outstanding achievement. If students are satisfied with their achievement, they may not be willing to work any harder to achieve more.

Results for grades 3 and 11. Five items dealing with the relationship of mathematics to oneself were given to both groups and are reported in table 10.3. Although two of the items were also included in the 7th- and 11th-grade set with the five-point scale, different response choices with a three-point scale were used here. Responses to identical item stems with different response choices are not comparable. The reader is invited to compare items C and K in table 10.2 with items E and A in table 10.3.

Overall, the elementary students expressed a much more positive attitude than the high school students. As many as 30 percent *more* said it was true that they were good at working with numbers and they were willing to work hard to do well in mathematics. Slightly over half of the students at both grade levels indicated that mathematics did not make them nervous. However, over half of the 11th-grade students indicated that mathematics was sometimes boring for them, whereas over half of the 3rd-grade students indicated that mathematics was never boring. Only 20 percent of the 3rd-grade students indicated that they did not like mathematics. (This item was not given to the older students.) These differences between elementary and high school students' perceptions of themselves and school mathematics are not surprising; vast social differences exist between the two groups and the younger group typically has a more enthusiastic attitude toward school.

Overall, students have a positive self-image of themselves as learners of mathematics, with the elementary school students expressing a more positive image than their senior high counterparts and the 7th-grade students being slightly more positive than the 11th-grade students.

Trends in views of mathematics and oneself. Common items administered in each of the last three NAEP Mathematics Assessments present a picture of trends in students' attitudes toward mathematics. Two of these items are presented in table 10.4. If students' stated desire to take additional mathe-

Table 10.4
Trends for Mathematics and Oneself (Ages 13 and 17)

		Percent Strongly Agree or Agree		
Statement		1978	1982	1986
I would like to take more mathematics.				
Nation:	13	50	47	43
	17	39	41	38
Male:	13	50	46	46
	17	42	43	40
Female:	13	49	48	40
	17	36	39	36
I am good at mathematics.				
Nation:	13	65	71	71
	17	54	58	61
Male:	13	70	76	74
	17	59	63	66
Female:	13	59	66	67
	17	49	53	55

matics courses is an indication of their enjoyment of the subject, it appeared that fewer 13-year-olds were enjoying mathematics in 1986 than in the 1978 and 1982 assessments. This drop was most pronounced for the 13-year-old females. The 17-year-olds, on the other hand, were relatively consistent over time in their desire to take more mathematics.

Students' confidence in their own mathematical ability increased significantly for both 13- and 17-year-olds from 1978 to 1982. However, the trend leveled off from 1982 to 1986. With the exception of 17-year-old females, whose reported confidence in mathematics steadily increased from 1978 to 1986, these patterns were generally the same for males and females. These changes in disposition toward mathematics did not correspond to the trends in average proficiency. From 1978 to 1982, 13-year-olds' performance increased, whereas they reported decreased enjoyment of, but increased confidence in, mathematics. For the 17-year-olds, performance improved significantly from 1982 to 1986 whereas they reported little change in their enjoyment of mathematics and only a moderate increase in their confidence in the subject.

MATHEMATICS AND SOCIETY

The items in this category dealt with students' perception of the usefulness of mathematics in everyday life and in their own career plans. In the 7th and 11th grades, five items with five-point Likert-type scales and "one yes-no-undecided" item were used. Only two three-point-scale items were administered to the 3rd-grade students.

The vast majority of the 7th- and 11th-grade students indicated that they thought mathematics had practical, everyday uses. Over three-fourths thought arithmetic was important for getting a good job. The response rate dropped to about 50 percent for the 11th-grade students when "mathematics such as algebra or geometry" was substituted for "arithmetic" in the previous statement. A much less dramatic change was obtained for the 7th-grade students. Fewer than half of the students at each of the three grade levels indicated that they expected to work in an area that requires mathematics. For the 3rd-grade students, the figure was only 40 percent. Also, only 40 percent of these younger students thought that most people use mathematics in their jobs. One-fourth were undecided about the use of mathematics. Students tended to view mathematics as important in society but less important for them personally. In a society where technology is rapidly increasing, such a narrow view of one's personal needs and future uses of mathematics could prove quite unfortunate. Students might fail to take today the mathematics needed later for the career of their choice and find their options seriously limited.

MATHEMATICS AS A DISCIPLINE

Mathematics as a school subject and a discipline. Items in this category dealt with perceptions of mathematics as being process oriented versus rule oriented and a dynamic rather than a static subject and with perceptions of mathematicians and mathematics as a formal discipline. These items were only given to the 7th- and 11th-grade students.

Both groups expressed quite similar views of mathematics. Over 80 percent of each group agreed that knowing how to solve a problem is as important as getting the solution and that knowing why an answer is correct is as important as getting the correct answer. However, approximately 80 percent of each group agreed that there is always a rule to follow in solving mathematics problems and that doing mathematics requires lots of practice in following rules. About half of each group reported that learning mathematics is mostly memorizing. Both groups of students viewed school mathematics as a subject in which both understanding and rules play important roles. About half of each group rejected the statement that exploring number patterns plays almost no part in mathematics and about two-thirds agreed that a mathematics problem can always be solved in different ways. But only about 40 percent agreed that a guess-and-check method can be used to solve a mathematics problem. Students appear to have an emerging view of mathematics as a process but also regard rules as still important. These results are presented in table 10.5.

Students had much less certain perceptions of mathematics as a discipline and of what a mathematician is like or does. Students did not have a clear view of mathematics as a good field for creative people or of mathematicians as people who work with ideas. Neither group seemed to have a strong view of mathematics as a cohesive discipline in which new discoveries are constantly being made. The older group, however, rejected negative statements about the nature of mathematics more frequently than the younger group, but such statements were rejected by fewer than 50 percent. The results are presented in table 10.6.

Students' views of mathematics and mathematicians were limited by their experience with school mathematics. It is not surprising that they expressed a somewhat confused and uncertain view of both. It is, however, encouraging to see that their view of school mathematics included an appreciation for understanding and process as well as product and that the older students with more experience in mathematics saw mathematics as a more dynamic, cohesive discipline.

Trends in views of mathematics as a discipline. Two items on the nature of mathematics as a discipline were included in the last three assessments. Smaller proportions of both males and females at ages 13 and 17 in 1986 than in 1978 agreed that mathematics helps a person to think logically.

Table 10.5
Mathematics as a Process

		Percent Responding[a]	
Statement	Grade	Agree	Disagree
A. Knowing how to solve a problem is as important as getting the solution.	7 11	88 92	4 3
B. Knowing why an answer is correct is as important as getting the correct answer.	7 11	82 89	6 4
C. Justifying the mathematical statements a person makes is an extremely important part of mathematics.	7 11	55 61	9 10
D. Mathematics helps a person to think logically.	7 11	64 71	12 9
E. There is always a rule to follow in solving mathematics problems.	7 11	83 81	8 9
F. Doing mathematics requires lots of practice in following rules.	7 11	78 86	8 5
G. Learning mathematics is mostly memorizing.	7 11	50 48	29 35
H. Exploring number patterns plays almost no part in mathematics.	7 11	16 15	48 55
I. A mathematical problem can always be solved in different ways.	7 11	65 60	15 20
J. Guess and check can be used to solve a mathematics problem.	7 11	42 40	33 37
K. Estimating is an important mathematics skill.	7 11	72 64	10 13

[a] Response rates on all items were above .95.

Table 10.6
Views of Mathematics and Mathematicians

		Percent Responding[a]	
Statement	Grade	Agree	Disagree
A. Creative people usually have trouble with mathematics.	7 11	23 16	39 50
B. Mathematicians work with symbols rather than ideas.	7 11	36 36	21 27
C. Mathematics is made up of unrelated topics.	7 11	23 19	41 50
D. New discoveries are seldom made in mathematics.	7 11	35 24	33 41

[a] Response rates for all items were above .95.

Over the three assessments, an increasing proportion of 17-year-olds agreed with the statement that new discoveries are seldom made in mathematics. In 1978, only about 20 percent held this view; by 1986, the

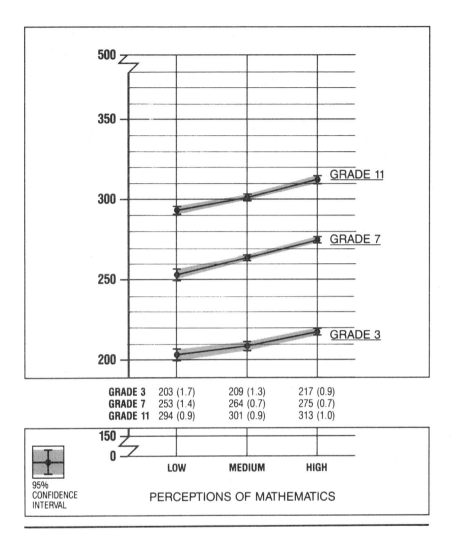

Fig. 10.1. Average mathematics proficiency by students' perceptions of mathematics, grades 3, 7, and 11: 1986. (Reprinted from *The Mathematics Report Card: Are We Measuring Up?* Princeton, N.J.: Educational Testing Service, 1988; p.93.)

proportion had risen to 35 percent. The views of 13-year-olds, however, did not change much over these three assessments.

Improvements in the levels of mathematics proficiency were not accompanied by improvements in students' views of mathematics as a dynamic discipline. Given that the improvements in student performance occurred predominantly on lower-level items, one might not expect a parallel increased appreciation and understanding of the nature of the discipline.

OVERALL DISPOSITION TOWARD MATHEMATICS AND MATHEMATICS PROFICIENCY

Although no attempt was made to construct attitude scales for the fourth mathematics assessment, a set of questions measuring students' perceptions of mathematics were compiled in a general background indicator for comparison with mathematics proficiency scores. Figure 10.1 (p. 115) shows the average mathematical proficiency scores for students in the 3rd, 7th, and 11th grades plotted against three dispositional levels—low, medium, and high. There appeared to be a positive relationship at all three grade levels between students' perceptions of mathematics and their proficiency in the subject. Students who have somewhat higher levels of mathematical proficiency are also likely to have a positive attitude toward the subject. Although there would appear to be a relationship between mathematics achievement and attitude, changes in attitudes and beliefs from one assessment to another do not appear to be directly related to trends in mathematics proficiency. As previously noted, 17-year-olds' mathematics proficiency increased from 1982 to 1986, but neither their enjoyment of the subject nor their perception of the nature of mathematics improved. Hence, the relationship between attitude and achievement would seem not to be a simple one.

11

WHAT CAN STUDENTS DO? (LEVELS OF MATHEMATICS PROFICIENCY FOR THE NATION AND DEMOGRAPHIC SUBGROUPS)

John A. Dossey, Ina V. S. Mullis, Mary M. Lindquist, and Donald L. Chambers

DEFINING LEVELS OF PROFICIENCY

RECENT calls for reform in mathematics education, a reaction to continuing poor performance in the 1970s, address the need to increase both average performance *and* the percentage of students reaching the higher ranges of proficiency. The recent improvement at age 9 as well as the signs of recovery at ages 13 and 17 are heartening and indicate some progress toward the first goal.

To describe more precisely the nature of mathematics performance and to document progress toward the second goal, that of helping more students reach the higher ranges of proficiency, NAEP has defined five levels of mathematics proficiency based on a retrospective analysis of the assessment results.

Using the range of student performance on the NAEP mathematics scale summarized in Chapter 1 of *The Mathematics Report Card* (Dossey et al 1988), five levels of mathematics proficiency were established: Level 150—Simple Arithmetic Facts, Level 200—Beginning Skills and Understanding, Level 250—Basic Operations and Beginning Problem Solving, Level 300—Moderately Complex Procedures and Reasoning, and Level 350—Multi-step Problem Solving and Algebra. Although proficiency levels above and below this range can theoretically be defined, few students in the NAEP sample performed at the extreme ends of the scale—that is, from 0 to 150 and from 350 to 500—and therefore any attempt to define other levels would have been highly speculative.

Benchmark questions were assigned to each proficiency level, based on the probability of correct responses. (Please refer to the Procedural Appendix in *The Mathematics Report Card* for a more elaborate discussion of the methods used to define proficiency levels.) Mathematics educators

---

This chapter is reprinted from *The Mathematics Report Card: Are We Measuring Up?* (June 1988) with the permission of the Educational Testing Service. For a more complete description of the analyses and scaling procedures, see the Procedural Appendix of the *Report Card*.

analyzed the empirically selected items and characterized the requisite skills held by students performing at each of the five levels of proficiency. Three factors appeared to affect performance: 1) the kind of mathematical operation students were asked to perform, 2) the type of numbers or number system involved, and 3) the problem situation. Students had less difficulty with basic operations, whole numbers, and straightforward problem settings. As the operations grew more involved and the problems moved out of the realm of whole numbers, performance levels decreased. Similarly, students had more difficulty with questions requiring the application of concepts, particularly in non-routine situations. Figure 11.1 provides a summary of the levels and their characteristic skills.

Levels of Mathematics Proficiency

Level 150—Simple Arithmetic Facts

Learners at this level know some basic addition and subtraction facts, and most can add two-digit numbers without regrouping. They recognize simple situations in which addition and subtraction apply. They also are developing rudimentary classification skills.

Level 200—Beginning Skills and Understanding

Learners at this level have considerable understanding of two-digit numbers. They can add two-digit numbers, but are still developing an ability to regroup in subtraction. They know some basic multiplication and division facts, recognize relations among coins, can read information from charts and graphs, and use simple measurement instruments. They are developing some reasoning skills.

Level 250—Basic Operations and Beginning Problem Solving

Learners at this level have an initial understanding of the four basic operations. They are able to apply whole number addition and subtraction skills to one-step word problems and money situations. In multiplication, they can find the product of a two-digit and a one-digit number. They can also compare information from graphs and charts, and are developing an ability to analyze simple logical relations.

Level 300—Moderately Complex Procedures and Reasoning

Learners at this level are developing an understanding of number systems. They can compute with decimals, simple fractions, and commonly encountered percents. They can identify geometric figures, measure lengths and angles, and calculate areas of rectangles. These students are also able to interpret simple inequalities, evaluate formulas, and solve simple linear equations. They can find averages, make decisions on information drawn from graphs, and use logical reasoning to solve problems. They are developing the skills to operate with signed numbers, exponents, and square roots.

Level 350—Multi-step Problem Solving and Algebra

Learners at this level can apply a range of reasoning skills to solve multi-step problems. They can solve routine problems involving fraction and percents, recognize properties of basic geometric figures, and work with exponents and square roots. They can solve a variety of two-step problems using variables, identify equivalent algebraic expressions, and solve linear equations and inequalities. They are developing an understanding of functions and coordinate systems.

Fig. 11.1

WHAT CAN STUDENTS DO? 119

Table 11.1 shows the percentage of students at ages 9, 13, and 17 who attained each level of proficiency in the 1978, 1982, and 1986 assessments. The highest mathematics levels attained across the three assessments by most students in each age group are highlighted, as are the 1986 percentages of 17-year-olds achieving the two highest proficiency levels.

Table 11.1
Trends for 9-, 13-, and 17-Year-Old Students
Percentage of Students at or Above the Five Proficiency Levels: 1978-1986

Proficiency Levels	Age	Assessment Year		
		1978	1982	1986
Level 150	9	96.5 (0.2)	97.2 (0.3)	**97.8 (0.2)**
Simple Arithmetic	13	99.8 (0.0)	99.9 (0.0)	100.0 (0.0)
Facts	17	100.0 (0.0)	100.0 (0.0)	100.0 (0.0)
Level 200	9	70.3 (0.9)*	71.5 (1.1)	**73.9 (1.1)**
Beginning Skills and	13	94.5 (0.4)*	**97.6 (0.4)**	**98.5 (0.2)**
Understandings	17	99.8 (0.0)	99.9 (0.1)	99.9 (0.1)
Level 250	9	19.4 (0.6)	18.7 (0.8)	20.8 (0.9)
Basic Operations and	13	64.9 (1.2)*	71.6 (1.2)	**73.1 (1.5)**
Beginning Problem Solving	17	92.1 (0.5)*	92.9 (0.5)*	**96.0 (0.4)**
Level 300	9	0.8 (0.1)	0.6 (0.1)	0.6 (0.2)
Moderately Complex	13	17.9 (0.7)	17.8 (0.9)	15.9 (1.0)
Procedures and Reasoning	17	51.4 (1.1)	48.3 (1.2)	**51.1 (1.2)**
Level 350	9	0.0 (0.0)	0.0 (0.0)	0.0 (0.0)
Multi-step Problem	13	0.9 (0.2)	0.5 (0.1)	0.4 (0.1)
Solving and Algebra	17	7.4 (0.4)	5.4 (0.4)	**6.4 (0.4)**

*Statistically significant difference from 1986 at the .05 level. (No significance test is reported when the proportion of students is either >95.0 or <5.0.) Jackknifed standard errors are presented in parentheses.

NATIONAL TRENDS IN LEVELS OF MATHEMATICS PROFICIENCY

Level 150: Simple Arithmetic Facts   1986

Age 9	Age 13	Age 17
97.8	100.0	100.0

Students performing at or above Level 150 are able to perform elementary addition and subtraction; however, their ability to apply these simple arithmetic procedures is likely to be quite constrained. Two sample items associated with Level 150 performance are provided in figure 11.2.

In 1986, as in the two previous assessments, virtually all students in each of the three age groups performed at or above Level 150. The results of the 1986 assessment indicate that American educators have been largely successful in their efforts to teach basic arithmetic skills to students in the initial grades.

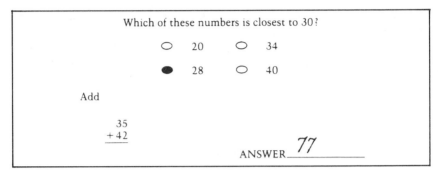

Fig. 11.2

Level 200: Beginning Skills and Understanding **1986**

Age 9	Age 13	Age 17
73.8	98.5	99.9

Students performing at or above Level 200 are developing a greater range and depth of basic mathematical skills however, their use of these skills is still imperfect and relatively inflexible. It can be inferred that learners at this level would have difficulty with reasoning that requires more than simple numerical computation. Seven sample items representative of Level 200 performance are provided in figure 11.3.

Virtually all 13- and 17-year-olds, and slightly less than three-quarters of the 9-year-olds, performed at or above Level 200 in the 1986 assessment. This represented significantly improved performance at both ages 9 and 13 between 1978 and 1986, indicating a rise in the proportion of students who have mastered low-level mathematical skills and knowledge.

Although these findings are generally positive, it must still be recognized that 26 percent of 9-year-olds who have not reached Level 200 constitute approximately 700,000 boys and girls in the third and fourth grades who have not yet acquired an understanding of rudimentary mathematical skills and concepts.

Level 250: Basic Operations **1986**
and Beginning Problem Solving

Age 9	Age 13	Age 17
20.8	73.1	96.0

Students performing at or beyond Level 250 on the proficiency scale have developed a surface understanding of the four basic operations, and

Subtract

$39 - 26 = \underline{13}$

$79 - 45 = \underline{34}$

Each bag has 10 marbles in it. How many marbles are there in all?

- ○ 10
- ○ 15
- ○ 25
- ○ 140
- ● 150
- ○ 160
- ○ I don't know.

Which coins are the same amount of money as a quarter?

- ○ 2 dimes
- ● 3 nickels and 1 dime
- ○ 3 dimes
- ○ 4 nickels
- ○ I don't know.

Find the quotient.

5)15

ANSWER $\underline{3}$

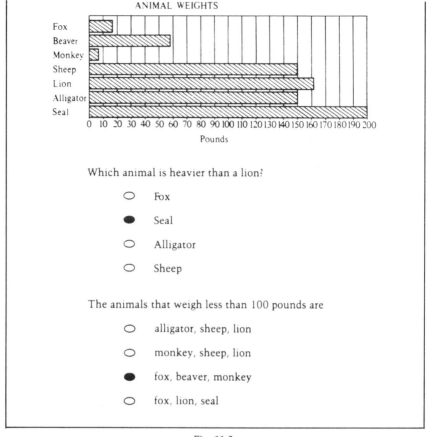

Fig. 11.3

are beginning to acquire more developed reasoning skills. A series of sample problems indicative of performance at Level 250 is provided in figure 11.4.

At Level 250, substantial differences in performance begin to appear across the age groups. Less than one-quarter of the 9-year-olds reached this level in any of the three most recent assessments. Given that basic computational skills are universally taught in the elementary grades, one would hope to see higher levels of proficiency at age 9 than have been found.

Although significantly more 13-year-olds performed at or above Level 250 in 1986 (73 percent) than in 1978 (65 percent), most of the gain occurred between 1978 and 1982, and the percentage of students achieving this level in either assessment is still quite low. Generalized to the nation as a whole, it is alarming that one-quarter of the seventh and

Which is worth the most?

- ○ 11 nickels
- ● 6 dimes
- ○ 1 half dollar
- ○ I don't know

Subtract

```
  604
- 207
```

ANSWER _397_

There are 10 airplanes on the ground. 6 take off and 4 more land. How many are on the ground then?

- ○ 4
- ● 8
- ○ 14
- ○ 20

At the store, the price of a carton of milk is 40¢, an apple is 25¢, and a box of crackers is 30¢. What is the cost of an apple and a carton of milk?

- ○ 55¢
- ● 65¢
- ○ 70¢
- ○ 95¢

BOXES OF FRUIT PICKED AT FARAWAY FARMS

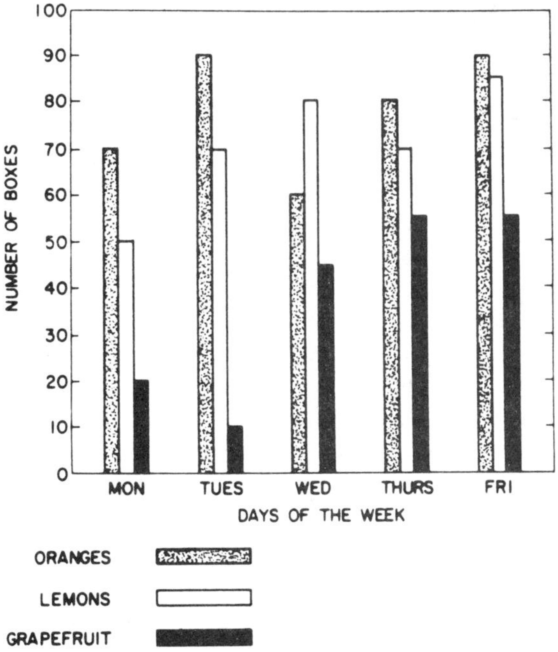

How many boxes of oranges, lemons, and grapefruit were picked on Tuesday?

- ○ 10
- ○ 90
- ● 170
- ○ 400
- ○ 940
- ○ 1700
- ○ I don't know.

> Find the product
>
> $$\begin{array}{r} 21 \\ \times\ 3 \\ \hline \end{array}$$
>
> ANSWER __63__
>
> Sam has 68 baseball cards. Juanita has 127. Which number sentence could be used to find how many more cards Juanita has than Sam?
>
> ● $127 - 68 = \square$
>
> ○ $127 + \square = 68$
>
> ○ $68 - \square = 127$
>
> ○ $68 + 127 = \square$
>
> ○ I don't know.

Fig. 11.4

eighth graders—amounting to more than three-quarters of a million students—may not possess the skills in whole-number computation necessary to perform many everyday tasks. The percentage of 17-year-olds performing at or above Level 250 also increased, from about 92 percent in both 1978 and 1982 to 96 percent in 1986. Although nearly all the high school students demonstrated proficiency in basic operations and beginning problem solving, the 4 percent of in-school 17-year-olds who did not reach this level would seem to be at a considerable disadvantage as adults—as would a presumably large proportion of their peers who have dropped out of school.

Level 300: Moderately Complex Procedures and Reasoning

1986		
Age 9	Age 13	Age 17
0.6	15.9	51.1

Students performing at or above Level 300 demonstrate more sophisticated numerical reasoning, and are beginning to draw from a wider range of mathematical skill areas, including algebra and geometry. A set of sample items representative of Level 300 performance is provided in figure 11.5.

In 1986, less than 1 percent of the 9-year-olds, 16 percent of the 13-year-olds, and 51 percent of the 17-year-olds were able to perform at or above this level. Further, at age 13, this reflects a decrease from the 1978 and 1982 assessments.

Which of the following is true about 87% of 10?

- ○ It is greater than 10.
- ● It is less than 10.
- ○ It is equal to 10.
- ○ Can't tell.
- ○ I don't know.

If 7x + 4 = 5x + 8, then x =

- ○ 1
- ● 2
- ○ 4
- ○ 6

What is the area of this rectangle?

- ○ 4 square cm
- ○ 6 square cm
- ○ 10 square cm
- ○ 20 square cm
- ● 24 square cm
- ○ I don't know.

6 cm
4 cm

Refer to the following graph. This graph shows how far a typical car travels after the brakes are applied.

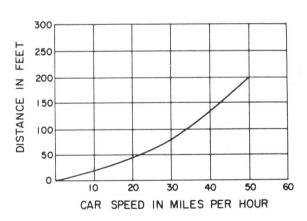

A car is traveling 55 miles per hour. About how far will it travel after applying the brakes?

- ○ 25 feet
- ○ 200 feet
- ● 240 feet
- ○ 350 feet
- ○ I don't know.

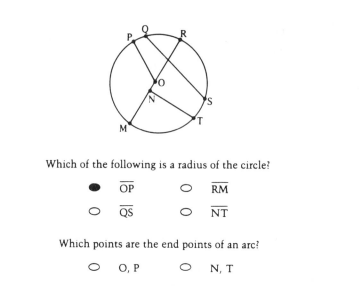

Fig. 11.5

Although the knowledge and problem-solving skills required to answer items at Level 300 are too advanced for 9-year-olds, it is troubling that more 13- and 17-year-olds have not attained this level of performance. Given that students are exposed to many of these topics in middle and junior high school, one would expect to see a higher percentage of students at age 13 and particularly at age 17 demonstrating success at this level of proficiency. The finding seems to lend support to recent calls for more challenging curriculum in the middle and upper grades.

Educating Americans for the 21st Century recommended that *all* secondary school students should achieve a variety of mathematics outcomes, including an understanding of the logic behind algebraic manipulations, a knowledge of two- and three-dimensional figures and their properties, and some more advanced objectives.[1] These recommendations were made to raise American secondary students' achievement so that it would be the best in the world by 1995. The performance data for Level 300 items in the NAEP mathematics assessment indicate that we have a great distance to go before our students achieve the levels defined by these recommendations.

---

[1] *Educating Americans for the 21st Century: A Plan of Action for Improving Mathematics, Science and Technology Education for All American Elementary and Secondary Students So That Their Achievement Is the Best in the World by 1995.* A Report to the American People and the National Science Board, the National Science Board Commission on Precollege Education in Mathematics, Science, and Technology, 1983.

WHAT CAN STUDENTS DO?

Further, the fact that nearly half of the 17-year-olds do not have mathematical skills beyond basic computation with whole numbers has serious implications. With such limited mathematical abilities, these students nearing graduation are unlikely to be able to match mathematical tools to the demands of various problem situations that permeate life and work.

Level 350: Multi-step Problem Solving and Algebra

1986

Age 9	Age 13	Age 17
0.0	0.4	6.4

Students performing at Level 350 demonstrate the capacity to apply mathematical operations in a variety of problem settings. A set of sample items representative of this level of performance is provided in figure 11.6.

R	S	40
35	25	15
T	V	W

In the figure above, R, S, T, V, and W represent numbers. The figure is called a magic square because adding the numbers in any row or column or diagonal results in the same sum. What is the value of R?

● 30 ○ 50
○ 40 ○ Can't tell

Suppose you have 10 coins and have at least one each of a quarter, a dime, a nickel, and a penny. What is the <u>least</u> amount of money you could have?

○ 41¢ ○ 50¢
● 47¢ ○ 82¢

If $f(x) = x^3 - x^2 + x - 4$, what is $f(-3)$?

● -43 ○ -1
○ -37 ○ 17

Christine borrowed $850 for one year from the Friendly Finance Company. If she paid 12% simple interest on the loan, what was the total amount she repaid?

ANSWER $952

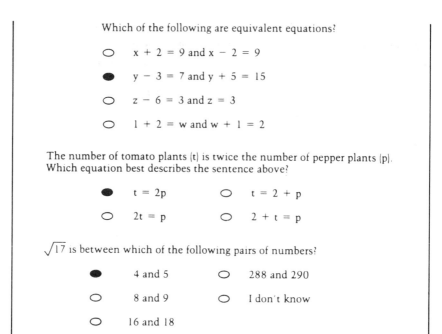

Fig. 11.6

Virtually no 9- or 13-year-olds and only a small proportion of 17-year-olds (6 percent) attained Level 350 performance in the 1986 assessment. Additionally troublesome is the fact that the percentage of students achieving at this level has remained essentially constant since 1978. At a time when mathematical and scientific skills are in high demand in the labor force, few students in their latter years of high school have mastered the fundamentals needed to perform more advanced mathematical operations.

Levels of Mathematics Proficiency for Demographic Subgroups

That most students fail to reach the higher levels of mathematics proficiency is sufficient cause for educators' concern. But how are special populations faring? A comparison of the mathematical abilities of various demographic subgroups with each other and with the nation as a whole offers a way to study variations in performance across subpopulations of interest. The populations of particular interest in this report are those distinguished by race, gender and region. (See the Procedural Appendix in *The Mathematics Report Card* for definitions.)

Levels of Proficiency by Race/Ethnicity

Essentially all students at ages 9, 13, and 17 performed at or above Level 150 in the 1986 assessment. However, even as early as age 9, there was slight variation across racial/ethnic groups in the percentage of students attaining this lowest level of proficiency. As illustrated in figure 11.7, a smaller percentage of Black and Hispanic 9-year-olds performed at Level 150 than did White students in this age group.

At all higher levels of proficiency, as well, White students consistently outperformed Hispanic students, and Hispanic students consistently outperformed Black students (with the single exception of 17-year-olds performing at Level 200). Disparities were especially striking among 9-year-olds at Level 200, among 13-year-olds at Level 250, and among 17-year-olds at Level 300. Thus, as age and level of proficiency increased, so did the performance gaps between racial/ethnic subpopulations.

Although these findings are discouraging, trends in levels of mathematics proficiency indicate considerable progress over the last eight years for racial/ethnic minorities. Unfortunately, as in the trends for the population at large, most of the increases occurred in the lower range of proficiency, primarily at Levels 200 and 250. (Please refer to Data Appendix for trend data.)

Levels of Proficiency by Gender

Within each age group, roughly the same percentages of males and females had mathematics proficiency at Level 150 or 200, as depicted in figure 11.8. However, variations began to appear among 13- and 17-year-olds who achieved at or beyond Level 250. Differences were particularly evident among 13-year-olds at Level 300, and among 17-year-olds at Levels 300 and 350, with more males achieving these higher levels than females.

Trends over time indicate that an increased proportion of both males and females attained Levels 150 to 250 from 1978 to 1986, and that the small gender performance gaps that existed at these levels in the 1978 and 1982 assessments have been further diminished. In the same time period, the proportions of both males and females achieving at Levels 300 and 350 have increased only slightly, and while there are still fewer females than males at both levels of proficiency, the performance gaps between the genders have not changed considerably.

Levels of Proficiency by Region

In examining the 1986 assessment results by region, it appears that differences across the four regions—Northeast, Southeast, Central, and West—are greatest at the upper levels of proficiency, primarily at Levels

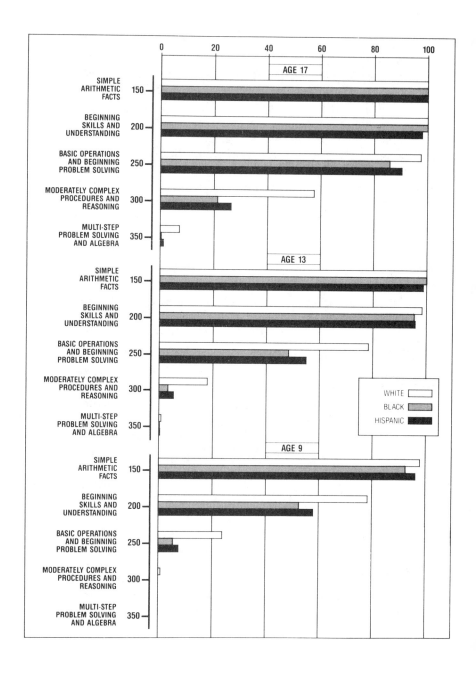

Fig. 11.7. Levels of mathematics proficiency: Percent at or above anchor points by race/ethnicity, 1986.

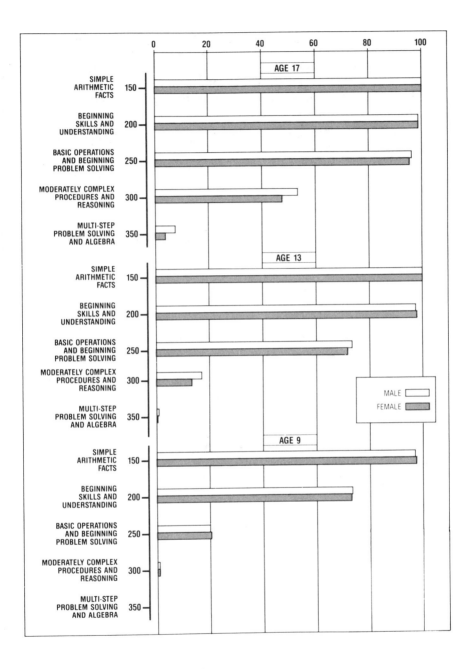

Fig. 11.8. Levels of mathematics proficiency: Percent at or above anchor points by gender, 1986.

250, 300, and 350. Figure 11.9 provides a comparative view of the levels of proficiency attained by students from each of these regions.

Although students from the Northeast, Central, and West regions were generally more likely than students from the Southeast region to attain Levels 200 to 350, trend data indicate that the latter group has made considerable progress since the 1978 assessment. For students from the Southeast, significant increases were evident from 1978 to 1986 in the percentage of both 9- and 13-year-olds performing at Level 200, and in the percentage of 17-year-olds performing at Level 250.

SUMMARY

As documented in Chapter 1 of *The Mathematics Report Card*, the mathematical performance of students at ages 9, 13, and 17 has improved somewhat over the past eight years, yet a closer look at levels of proficiency indicates that most of the progress has occurred in the domain of lower-order skills.

The more detailed account of what students can and cannot do at each level of proficiency gives further weight to educators' concerns that many students lack skills commonly thought to be mastered at the elementary, middle, and high school levels. It appears that the discrepancy between students' expected and actual mathematics performance begins early on in schooling, and increases as they move into the upper grades. One would expect a majority of 9-year-olds (primarily fourth graders) to have mastered basic mathematical operations and beginning problem solving (at Level 250), as these skills are usually taught in elementary school. The fact that only 21 percent of the 9-year-olds attained this level in the 1986 assessment and that one-quarter of them failed to demonstrate even beginning skills and understanding (Level 200) suggests that reform in the mathematics curriculum may be warranted from the earliest grades.

At age 13, the discrepancy between students' expected and actual proficiency is larger still. Moderately complex mathematical procedures and reasoning (at Level 300) generally are embedded throughout the middle and junior high school curriculum, yet in 1986 only 16 percent of the students assessed at age 13 demonstrated a grasp of these skills. It seems likely that children who did not receive a strong mathematics foundation in the elementary grades have increasing difficulty in the subject through their middle years of schooling, as more difficult operations and concepts are introduced.

The discrepancy between expected and actual performance grows even more pronounced among 17-year-olds, of whom only 6 percent in the 1986 assessment displayed abilities in multi-step problem solving and algebra (at Level 350). Only about half of the 17-year-olds demonstrated even a moderately complex understanding of mathematics, as

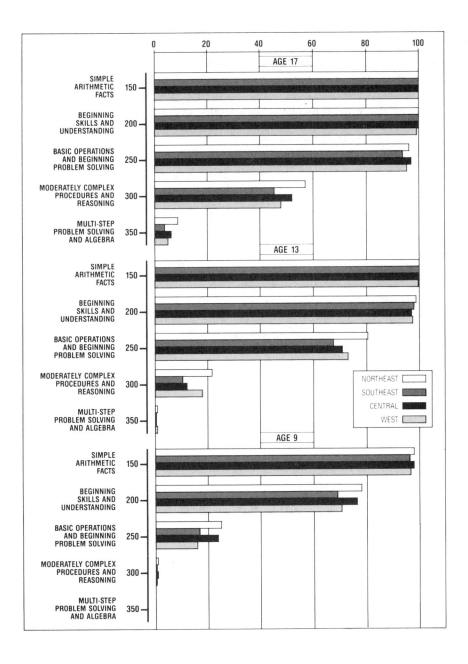

Fig. 11.9. Levels of mathematics proficiency: Percent at or above anchor points by region, 1986.

exemplified by Level 300 performance. Translated into population figures, nearly 1.5 million 17-year-old students across the nation appear scarcely able to perform the kinds of numerical operations that will likely be required of them in future life and work settings.

These concerns are heightened by the fact that the 17-year-olds sampled in the fourth mathematics assessment did not include the 14 percent of the cohort who have already dropped out of school by junior year.[2] It is quite likely that the mathematical proficiency of these absent students is considerably lower; hence the 1.5-million figure quoted previously would be optimistic for the entire 17-year-old population.

For subpopulations whose mathematics performance has tended to lie below national averages in NAEP assessments—including Black and Hispanic students and residents of the Southeast—the discrepancy between expected and actual performance for all age groups remains even larger than that for the nation as a whole, despite considerable gains in recent years.

The levels of proficiency exhibited by American students, particularly in the higher age groups, are likely to be inadequate for the demands of the times. A nation that wants to continue to reap the benefits of modern technology and to compete in the future global economy depends on the skills of the young, and it appears that our students are ill-prepared to meet these challenges.

---

[2] *The Condition of Education, 1984 Edition.* National Center for Education Statistics (NCES), 1985.

12

MINORITY DIFFERENCES IN MATHEMATICS

Martin L. Johnson

THIS NAEP administration allowed a comparison of performance among white, black, and Hispanic students. Although other ethnic groups were included in the assessment, the sample sizes were insufficient for separate analyses. In past assessments, black and Hispanic students have performed consistently below white students in virtually every content category of the assessment. Unfortunately, this trend in performance continues. In this chapter specific items are discussed that highlight these differences in performance. The reader is reminded that previous chapters have discussed performance of all students across content areas, and no attempt is made to duplicate those findings in detail. Instead, this chapter discusses overall performance of minority students across content categories and the responses of minority students to the attitude items.

HIGHLIGHTS

• Blacks and Hispanics made substantial gains from 1982 to 1986 with the gains for blacks being significant across all grade levels at the .05 level.

• Performance gains of black and Hispanic students were greater at the 200 and 250 mathematics proficiency levels.

• Hispanics made substantial gains in enrollment in advanced level courses since 1978; blacks made modest gains.

• Across all content domains, black and Hispanic students performed at comparable levels, however, performance of both groups was consistently below that of white students on all items except those testing the lowest level of whole number computation.

• All minority students reported very positive attitudes toward mathematics, held high expectations toward their performance in mathematics, and reported that their parents held equally high expectations of their performance.

OVERALL PERFORMANCE TRENDS

Blacks made significant (at the .05 level) gains at all age levels from 1982 to 1986. These gains continue a pattern extending back to the 1978 assessment; at age 13 the gains were substantial from 1978 to 1982 and from 1982 to 1986. Tables 12.1 and 12.2 present data on mean mathematics proficiency and the percentage of students at or above the five mathematics proficiency levels discussed in chapter 1. Although Hispanic students also made gains, most were not significant at the .05 level. Table 12.2 shows that at age 13, the gains for blacks and Hispanics at the 200 level essentially wiped out the

Table 12.1
Mean Mathematics Proficiency (Weighted Mathematics Proficiency Means and Jackknifed Standard Errors)

	White	Black	Hispanic
Age 9			
1977–78	224.1 (0.9)	192.4 (1.1)*	202.9 (2.3)
1981–82	224.0 (1.1)	194.9 (1.6)*	204.0 (1.3)
1985–86	226.9 (1.1)	201.6 (1.6)	205.4 (2.1)
Age 13			
1977–78	271.6 (0.9)	229.6 (1.9)*	238.0 (2.2)
1981–82	274.4 (1.0)	240.4 (1.6)*	252.4 (1.6)
1985–86	273.6 (1.3)	249.2 (2.3)	254.3 (2.9)
Age 17			
1977–78	305.9 (0.9)*	268.4 (1.3)*	276.3 (2.2)
1981–82	303.7 (0.9)	271.8 (1.3)*	276.7 (2.0)
1985–86	307.5 (1.0)	278.6 (2.1)	283.1 (2.9)

*Significant difference from 1986 at the .05 level.
(Source: *The Mathematics Report Card*, p. 138)

Table 12.2
Percentage of Students at or above the Five Mathematics Proficiency Levels

Levels		1977–78			1981–82			1985–86		
		W	B	H	W	B	H	W	B	H
150	Age 9	98	88	93	98	90	95	99	93	96
200	Age 9	76	43	54	77	47	55	79	53	59
	Age 13	98	80	86	99	89	96	99	96	96
	Age 17	100	99	99	100	100	100	100	100	99
250	Age 9	26	4	11	22	5	9	25	5	8
	Age 13	73	29	36	79	38	54	79	49	55
	Age 17	96	70	77	96	75	81	98	86	91
300	Age 13	21	2	3	21	3	6	19	4	5
	Age 17	57	18	22	55	17	21	58	22	27

Key: W: white; B: black; H: Hispanic
(All numbers rounded to nearest whole number)
(Source: *The Mathematics Report Card*, p. 139–41)

lowest levels of deficiency found in earlier assessments. About 80 percent of the black students were at the 200 level in 1978; by 1986 this figure had risen to over 95 percent. At the 250 level, the gains were equally impressive. The number of blacks and Hispanics attaining the 250 level almost doubled from 1978 to 1986 (from 29% to 49% for blacks and from 36% to 50% for Hispanics). However, the gains were relatively small at the 300 and 350 levels. It appears that we have been fairly successful in improving the performance of many students who were previously unsuccessful but only at the lower end of the scale.

Blacks and Hispanics at age 17 also made great improvement at the 250 level (table 12.2). Over this eight-year span blacks improved from 70 percent to 86 percent and Hispanics improved from 77 percent to 91 percent. Neither group at age 17 showed any significant gains at levels above 250, indicating that instruction has been effective for improving performance on lower-level items but not effective for items involving complex reasoning skills and problem solving.

Table 12.3 presents data on course-taking behavior from 1978 to 1986. Gains in enrollment in advanced level courses have been substantial, particularly among Hispanics. Enrollment in Algebra 1 increased from 57 percent to 74 percent, enrollment in geometry increased from 38 percent to 50 percent, and enrollment in Algebra 2 increased from 26 percent to 34 percent. Enrollment of blacks also increased slightly.

The trends over the eight-year span are positive and suggest that the performance of black and Hispanic students is increasing and that both groups are enrolling in advanced mathematics courses with increasing frequency. It remains clear, however, that the performance of black and Hispanic students remains at a lower level than that of white students.

Table 12.3
High School Course Enrollment (Percent of 17-Year-Olds Who Have Taken a Given Course)

Course	White	Black	Hispanic
Algebra 1			
1978	79	62	57
1982	76	63	60
1986	83	68	74
Geometry			
1978	62	43	38
1982	61	43	39
1986	66	50	50
Algebra 2			
1978	45	32	26
1982	46	33	27
1986	49	34	34

CONTENT

The section of the assessment that focused on fundamental methods of mathematics contained items requiring different methods of solution including students' patterns of reasoning. This area gives a measure of how the mathematics curriculum and mathematics teaching are developing students who can reason about mathematical problems and situations. Many of the questions focused on logical reasoning, in the form of "if a then b," or on transitive reasoning. A sample item, given in both grades 3 and 7, is presented in table 12.4. Black and Hispanic students performed well below white students at each grade level and below the 33 percent level at grade 7. The high rate of response for the incorrect answer "blue" indicates that students disregarded the fact that the blue car is next to the red. Although black and Hispanic students in grade 7 performed better on this item than students in grade 3, the same percentage of the 7th-grade students as 3rd-grade students chose the incorrect response, indicating very little growth in the type of logical reasoning measured by this task.

Table 12.4
Logical Reasoning (Grades 3 and 7)

	Percent Responding					
	Grade 3			Grade 7		
Item	White	Black	Hispanic	White	Black	Hispanic
Four cars wait in a single line at a traffic light. The red car is first in line. The blue car is next in line. The green car is between the blue car and the white car. Which color car is at the end of the line?						
Red	5	7	10	2	2	3
Blue	54	63	56	45	63	59
Green	9	11	13	3	5	8
White*	32	20	21	49	30	31

*Indicates correct response.

Table 12.5 shows other forms of logical reasoning items and students' performance on typical items. The performance in grade 3 was lower on item A for blacks and Hispanics than for whites but about the same in grade 7. Performance was low on item B for all groups of students. Item B involved the use of deductive reasoning often needed in algebra and geometry. The performance of black and Hispanic students also lagged behind that of whites on item C. This item, presented in table 2.2, involved reasoning from a mathematical diagram.

MINORITY DIFFERENCES IN MATHEMATICS

Table 12.5
Logical Reasoning

	Percent Correct								
	Grade 3			Grade 7			Grade 11		
Item description	W	B	H	W	B	H	W	B	H
A. Figure X is smaller than figure Y, and figure Z is larger than figure Y. (Choose figure X from three given figures.)	67	44	49	85	79	77	—	—	—
B. Given a true if-then statement. (Choose a statement that could not be true.)				48	37	34	54	39	40
C. Reason from a mathematical diagram (see table 2.2).							58	40	40

W = white; B = black; H = Hispanic

This overall pattern of performance among different groups of students was also evident in the item shown in table 12.6, which involved reasoning about numbers and number relationships. As table 12.6 shows, only 38 and 41 percent of black and Hispanic students, respectively, in grade 11 gave correct answers. One wonders whether this performance reflects a lack of instruction about logical reasoning about mathematical relationships or, if instruction was given, whether this performance is an accurate assessment of what was learned.

Table 12.6
Reasoning about Numbers (Grade 11)

	Percent Responding		
Item	White	Black	Hispanic
What number am I?			
(i) If you add my digits you get 15.			
(ii) My tens digit is one more than my hundreds digit.			
(iii) My ones digit is one more than my tens digit.			
834	10	16	13
780	15	26	28
678	11	19	18
456*	65	38	41

*Indicates correct response.

The items in table 12.4-12.6 are representative of those assessing the broad area of fundamental methods of mathematics. Overall, the performance of black and Hispanic students lagged behind that of whites. Blacks and Hispanics seemed to omit (or not reach) more items than other groups, making it impossible to interpret their responses to many interesting items.

The ability to read graphs and interpret data was assessed by many items in the data organization and interpretation section. Table 12.7 shows students' performance in interpreting different graphical representations.

Table 12.7
Reading Tables and Graphs

	Percent Correct								
	Grade 3			Grade 7			Grade 11		
Item	W	B	H	W	B	H	W	B	H
A. Tally Chart	70	46	51	—	—	—	—	—	—
B. Circle Graph									
Simple question	77	62	59	97	94	91	—	—	—
Complex question				73	68	65	—	—	—
C. Bar Graph									
Simple question	71	51	54	88	85	78	—	—	—
Complex question				70	50	46	78	62	50
D. Line Graph									
Simple question				48	26	32	69	48	52
Complex question							87	67	69

W = white; B = black; H = Hispanic

Fewer than 50 percent of black students and only 51 percent of Hispanics interpreted a tally graph correctly in grade 3. Circle graphs were easier than either bar or line graphs; however, at grade 11 only 67 percent of blacks and 69 percent of Hispanics could answer a complex question about a line graph. When students were asked to read and interpret a line graph in a problem-solving context, as in table 12.8, performance fell drastically. One-fourth of the black students and one-sixth of the Hispanic students at grade 7 responded "I don't know" to this item. This may have reflected the small amount of instruction they received on this topic by grade 7, yet graphing is included in the prescribed K–7 mathematics curriculum of every school system. At grade 11, approximately 50 percent of each minority group responded correctly, indicating some improvement.

Overall, it is clear that minority students experienced much difficulty in reading graphs and interpreting information presented in graphical form. This area was extremely troublesome for black and Hispanic students—they reached the 90 percent level of performance on only one of the twenty-one items at grade 7 and only one of the twenty items at grade 11.

Performance on the measurement and geometry items was extremely

MINORITY DIFFERENCES IN MATHEMATICS

Table 12.8
Interpretation of Graphical Data (Grades 7 and 11)

The graph shows how far a typical car travels after its brakes are applied. A car is traveling at 30 miles per hour. About how far will it travel after applying the brakes?

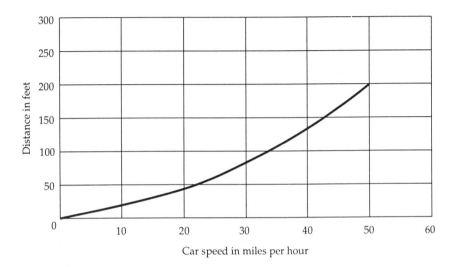

	Percent Responding					
	Grade 7			Grade 11		
Possible Responses	White	Black	Hispanic	White	Black	Hispanic
16 feet	14	18	17	4	11	7
45 feet	6	12	10	3	7	10
65 feet	18	17	20	19	19	19
80 feet*	48	26	33	69	48	52
140 feet	4	3	4	2	4	4
I don't know	11	25	16	3	11	8

*Indicates correct response.

weak among all students. At grade 3, a large number of items was omitted, indicating that little measurement has been taught by grade 3 or that students did not attempt the items. At grade 7, a large number of items were given, and performance decreased as the concepts changed from perimeter to area to volume. Performance of black and Hispanic students continued to lag behind that of white and other minority students at each grade level. Table 12.9 presents descriptions of some items from this part of the assessment.

Table 12.9
Measurement and Geometry (Grades 7 and 11)

	Percent Correct					
	Grade 7			Grade 11		
Item description	White	Black	Hispanic	White	Black	Hispanic
A. Choose the largest metric length unit.	66	48	51	75	54	65
B. Measure object to nearest quarter inch.	83	67	68	—	—	—
C. Find the area of a rectangle.	50	33	31	74	51	52
D. Identify a diameter of a circle.	77	55	65	91	76	86
E. Identify a radius of a circle.	47	32	34	70	46	57
F. Choose the name of the figure suggested by people standing the same distance from a given person.	39	22	24	70	35	43

Item A asks for the relationship among metric units for linear measure. Fifty-four percent of the black students and 64 percent of the Hispanic students answered correctly. The unit "meter" was a very popular choice, suggesting that units larger than a meter were unfamiliar to many students. Black and Hispanic performance was well below that of white students on item B as well. When the facts that this was the performance of students in grade 7, that the content of the items is basic to linear measurement, and that it was probably first taught in elementary grades are considered, serious questions must be raised about what is being taught under the heading of "measurement" in the middle and high school.

Items C–E assessed basic geometry concepts of area, diameter, and radius. Black and Hispanic students in grade 11 were able to identify the diameter of a circle but performed poorly on the other items. On item F fewer than half of the black and Hispanic students at grade 11 could determine the figure suggested by four people standing the same distance from a given fifth person. It was clear from the patterns of response in both grades 7 and 11 that students focused either on three of the people and chose the triangle as the response or on the four people and chose the rectangle as the response. Perhaps students had not been exposed to this way of defining a circle; however, on a companion item asking students to choose from a set of drawings the figure having all its points the same distance from a point p, 41 percent of the black students and 53 percent of the Hispanic students correctly chose the circle. This indicates that if students are given a drawing and asked if it satisfies specific criteria, performance improves slightly.

The performance discussed here is representative of black and Hispanic students throughout the section assessing geometry knowledge and understanding. There were two notable exceptions: minority students in grade 3 can find the mirror image of a figure by paperfolding and students in grade 7 can correctly identify parallel lines. Overall, minority performance in geometry was very low.

Minority students performed comparably with other students at grade 3 in the area of relations, functions, and algebraic expressions. Overall, however, grade 3 students answered few questions in this section. Grade 3 children performed very well with expressions such as $3 + \_\_ = 7$ and $3 \times \_\_ = 21$. Sixty-seven percent of black and 71 percent of Hispanic students answered the latter item correctly. However, at the higher grades and on items that contained algebraic expressions, the performance of minority students showed a drastic decline. One such item is given in table 12.10. Although the performance was low for all students on this item, only 30 percent of the black and 32 percent of the Hispanic students in grade 11 answered it correctly. Of interest is the large percentage of all students who chose $2p = t$ as the correct response. Even in grade 11 the response $2p = t$ was chosen most often by black and Hispanic students. This pattern of performance is consistent with performance on previous items assessing logical reasoning and mathematical relationships. It was also evident that throughout the algebra problems, the omission rate of black children was around 44 percent. This is very unsettling given that other data indicated that the number of black and Hispanic students enrolling in algebra classes had increased since the last assessment.

The remaining content category in the assessment covered numbers and operations, both knowledge and skill and higher-level operations. The

Table 12.10
Algebraic Expressions (Grades 7 and 11)

	Percent responding					
	Grade 7			Grade 11		
	White	Black	Hispanic	White	Black	Hispanic
The number of tomato plants (t) is twice the number of pepper plants (p). Which equation best describes the sentence above?						
$t = 2p$*	26	19	21	42	30	32
$2t = p$	36	38	33	48	50	46
$t = 2 + p$	25	30	34	6	14	12
$2 + t = p$	14	13	12	4	6	10

*Indicates correct response.

performance on this category is discussed in detail in chapter 8. As for the performance of black and Hispanic students, performance differences were small on whole number computation items. Second, on computational items involving rational numbers or percents, the performance gap between black and Hispanic students and white students was large. For example, on the item requiring subtraction of fractions, 35 percent of the white students performed correctly at grade 7 but only 16 and 13 percent of the black and Hispanic students, respectively, did the exercise correctly. Third, in word problem situations and in the application of number and operations to problem solving, the gap in performance remains large between black and Hispanic students and white students. Table 12.11 contains performance data for items being discussed.

Table 12.11
Number and Operations (Grades 7 and 11)

	Percent Correct					
	Grade 7			Grade 11		
Item	White	Black	Hispanic	White	Black	Hispanic
Whole Number Computation						
A. 242 − 178	86	82	82	92	89	89
B. 504 − 306	86	76	81	90	86	90
C. 213 × 12	80	69	75	85	84	86
Rational Number Computation						
A. 3½ − 3⅓	58	36	34	73	59	60
B. 9⅛ − 2½	35	16	13	47	32	30
C. 4 × 2¼	61	41	44	73	58	54
D. 8% of 25	34	23	29	64	49	54
E. 7.2 × 2.5	66	51	55	69	56	65
Word Problems						
A. Two-step (−, ÷)	57	35	38	—	—	—
B. % of a number	41	22	34	72	52	54

The analyses of trend data across many NAEP assessments reveal that the performance of black and Hispanic students is improving. It appears, however, that the gains were made almost entirely in the area of computational skills. Although these gains are important, efforts must be made to increase student performance in other areas of the mathematical curriculum.

ATTITUDES

The overall purpose, structure, and results of the attitude items are discussed in chapter 10. Since it is generally believed that a positive relationship exists between one's attitude toward mathematics and one's performance, it is important to be aware of the attitudes of different ethnic groups as one factor that might influence their mathematics performance. Table 12.12 contains responses of black, Hispanic, and white students in grades 7 and 11 to items comparing mathematics with other school subjects on the dimensions of like-dislike, easy-hard, and important-not important. It is clear that all students in grade 7 liked mathematics, a majority thought it was easy, and almost all thought it was important. Although all the academic subjects were liked by over two-thirds of the black seventh-grade students, mathematics was liked best, receiving a positive response from 75 percent of these students. Mathematics and English were the preferences of the Hispanic seventh-grade students, with 62 percent liking these subjects. Mathematics and science were the preferred subjects of the white students. On the easy-hard dimension, mathematics was rated as easy by slightly more of the minority students in grade 7 than either social studies or science, by slightly fewer students than English.

Table 12.12
Mathematics and Other School Subjects

	Percent Responding								
	Like			Easy			Important		
Subject	B	H	W	B	H	W	B	H	W
Grade 7									
Mathematics	75	62	66	59	51	52	94	87	91
English	71	62	56	65	59	52	89	81	84
Science	71	58	67	53	37	44	74	58	72
Social science	65	49	56	54	43	44	78	68	65
Physical education	82	72	76	86	77	81	50	44	45
Grade 11									
Mathematics	72	59	61	44	41	39	93	90	85
English	83	64	64	66	49	50	94	88	86
Science	73	66	62	52	43	36	74	68	67
Social science	70	61	60	61	54	53	66	57	58
Physical education	75	79	72	91	88	84	42	50	39

B = black; H = Hispanic; W = white

This picture had changed by grade 11. Mathematics was no longer the best-liked academic subject. Although over 70 percent of the black students still liked mathematics, this figure is only slightly different from the percentage that liked social studies and science and lower than the percentage that liked English. Not much difference in the preferences for subjects was shown by the Hispanic 11th-grade students, but mathematics was liked least. White students preferred English but only slightly more than mathematics. Only 44, 41, and 39 percent of the black, Hispanic, and white students, respectively, indicated that they thought mathematics was easy, a drop of about 15, 10, and 13 percentage points, respectively, from the seventh-grade responses. Except for ratings of English among the Hispanic students, the other subjects did not show such a change in difficulty ratings from the seventh to the eleventh grade. English joined mathematics as the most important subjects as far as all students were concerned. Even with the dramatic change of percentages from grade 7 to 11, mathematics rated well in importance in comparison with other subjects.

Students' views of mathematics and of themselves are given in tables 12.13 and 12.14. The responses of third-grade minority students (table 12.13) indicates that they had a positive view of themselves as learners of mathematics in that they liked mathematics, felt they were good with numbers, and were willing to work hard. The most obvious conclusions to be drawn from data in table 12.14 are the following:

(a) Minority students in grade 11 are willing to work hard to do well in mathematics.

(b) Minority students perceive that their parents expect them to do well in mathematics.

(c) Minority students have high expectations of themselves in mathematics.

(d) Good grades are perceived as important.

(e) Students like to be challenged and feel good when they are able to solve mathematics problems on their own.

Table 12.13
Mathematics and Oneself (Grade 3)

		Percent Responding[a]		
Statement	Ethnic Group	True	Sometimes True	Not True
A. I usually understand what we are talking about in mathematics.	Black	49	39	12
	Hispanic	51	38	11
	White	56	33	10
B. I am good at working with numbers.	Black	66	29	5
	Hispanic	65	24	11
	White	65	30	5
C. Doing mathematics makes me nervous.	Black	24	28	48
	Hispanic	19	31	50
	White	17	26	57
D. Mathematics is boring to me.	Black	21	23	56
	Hispanic	25	19	56
	White	21	23	57
E. I am willing to work hard to do well in mathematics.	Black	79	15	6
	Hispanic	76	14	11
	White	83	12	4

[a] Response rates for all items were above 77%.

Table 12.14
Mathematics and Oneself (Grades 7 and 11)

Statement	Grade	Percent Agreeing		
		Black	Hispanic	White
A. I really want to do well in mathematics.	7	96	90	93
	11	95	87	82
B. My parents want me to do well in mathematics.	7	96	90	94
	11	91	80	84
C. I am willing to work hard to do well in mathematics.	7	89	82	84
	11	90	82	77
D. A good grade in mathematics is important to me.	7	92	87	89
	11	93	83	78
E. I enjoy mathematics	7	63	53	54
	11	61	56	47
F. I like to be challenged when I am given a difficult mathematics problem.	7	60	51	54
	11	60	45	48
G. I feel good when I solve a mathematics problem by myself.	7	90	81	84
	11	94	89	87
H. I am taking mathematics only because I have to.	7	33	39	34
	11	20	35	27
I. I would like to take more mathematics.	7	49	46	43
	11	42	46	38
J. I am good at mathematics.	7	58	50	61
	11	55	46	53
K. I usually understand what we are talking about in mathematics.	7	80	72	78
	11	74	65	67

The majority of students reported that they were not taking mathematics only because they had to, and about half of all students at grade 7 reported that they would like to take more mathematics. The responses reported in tables 12.13 and 12.14 are encouraging and challenge teachers to use these positive attitudes in the pursuit of higher levels of learning by these students.

Students in both grades 7 and 11 were asked to respond to a set of items involving mathematics as a process. The responses are given in table 12.15. The responses to most items indicate that minority students were aware that there is more to mathematics than just "getting an answer" and that many different procedures can be used in problem solving. It is, however, also clear from responses to items E and F that minority students still saw mathematics as "following rules" and as a subject requiring lots of practice. About the same percentage of white 11th-grade students as the minority students agreed that mathematics requires practice, but over 10 percent fewer white students than minority students thought that there is always a rule. When asked if mathematics has practical use, 80 percent, 85 percent, and 82 percent of the white, black, and Hispanic students, respectively,

Table 12.15
Mathematics as a Process (Grades 7 and 11)

		Percent Agreeing		
Statement	Grade	Black	Hispanic	White
A. Knowing how to solve a problem is as important as getting the solution.	7 11	87 93	84 90	89 91
B. Knowing why an answer is correct is as important as getting the correct answer.	7 11	80 93	74 90	84 88
C. There is always a rule to follow in solving mathematics problems.	7 11	88 93	84 91	81 79
D. Doing mathematics requires lots of practice in following rules.	7 11	82 86	76 84	78 86
E. A mathematical problem can be solved in different ways.	7 11	77 79	69 70	63 55

responded positively. Also, 59 percent, 68 percent, and 52 percent of the white, black, and Hispanic students, respectively, disagreed with the statement that you can get along without mathematics in daily life.

SUMMARY

The mathematics performance of black and Hispanic youth continues to lag behind the performance of white students. This situation continues to be a major concern of mathematics educators and policy makers at all levels. The small gains that have been measured over the last three national assessments are in computational skill, considered to be the least complex level of performance measured on this test. Computational performance notwithstanding, gains in other areas over the last three assessments are more difficult to find. Within the current assessment, the increase in performance from grade to grade on items other than simple ones was very slight among black and Hispanic students.

Contrary to popular belief, black and Hispanic students reported very positive attitudes towards mathematics. They and their parents had high expectations, and they were willing to work hard to achieve in mathematics. They realized the importance of mathematics in everyday life, and enrollment in higher mathematics classes has increased. It is a challenge to all to capitalize on these seemingly positive attitudes in helping these students to achieve higher levels of performance.

13

GENDER DIFFERENCES IN MATHEMATICS

Margaret R. Meyer

IN THE past decade, the National Assessment of Educational Progress has been an important source of data on gender differences in mathematics. The information it gathers about mathematics course taking, achievement, and attitudes for a large national sample of students enables us to monitor the relative progress of females and males. This chapter reports on the gender differences found in the 1986 mathematics assessment.

One benefit of the periodic assessments is that changes from one assessment to another can be analyzed; however, because of a change in the sampling process, comparisons between this assessment and those of previous years are limited. Instead of sampling by age (9-, 13-, and 17-year-olds) as was done in the previous three assessments, the 1986 assessment sampled by grade (grades 3, 7, and 11). For the most part, this method resulted in a sample with characteristics similar to those of past samples. However, care must be taken when results from the fourth assessment are compared with those of the previous three. In this chapter, when comparisons are made across assessments, students will be identified by their age rather than by grade.

MATHEMATICS COURSE TAKING

Table 13.1 shows the percentage of 17-year-old females and males who reported a specific course as their highest mathematics course taken. Data from the 1978 and 1982 assessments are included to show changes in course-taking patterns over the past decade. As with these previous assessments, very little difference was found between females and males in mathematics courses taken through Algebra 2. Differences favoring males emerged in the elective, or fourth-year, mathematics courses—males were much more likely to elect precalculus or calculus.

The NAEP also asked students about other elective mathematics courses. At age 17, males were more likely than females to elect a course on

Table 13.1
Changes in Percentages of Females and Males Taking Mathematics Courses, Age 17

Course	Percentages of 17-Year-Olds Reporting Highest Level of Course Taken	
	Males	Females
Algebra 1		
1978	15	18
1982	16	17
1986	17	18
Geometry		
1978	15	18
1982	13	15
1986	15	18
Algebra 2		
1978	38	37
1982	39	39
1986	39	40
Precalculus or Calculus		
1978	7	4
1982	6	5
1986	8	5

probability or statistics (5% versus 3%, respectively), and a full year of computer programming (18% versus 10%, respectively), although males and females were equally likely to elect half a year of computer programming (13% versus 14%, respectively). Males' higher enrollment in these elective courses may indicate a continuing trend for more males than females to enroll in college-level mathematics and pursue mathematics-related careers. It should be noted that these data were collected in the eleventh grade and, therefore, few students would have had a chance to take a fourth year of mathematics. However, the relative percentages still suggest a greater participation of males at the highest levels of mathematics.

GENDER DIFFERENCES IN ACHIEVEMENT

Proficiency Levels

Figure 13.1 shows the national trends in mathematics proficiency for 9-, 13-, and 17-year-old females and males over the last three assessments. The pattern of gender differences found in the 1978 and 1982 assessments is again apparent in the 1986 data. At age 9, the average proficiency levels for females and males were equivalent. From 1982 to 1986, males made a significant gain in proficiency, whereas the gain of females in that same time was minimal. By age 13, females had lost whatever small advantage they had had over males. In both 1982 and 1986, the 13-year-old males had slightly higher proficiency levels than their female peers. At age 17, the

GENDER DIFFERENCES IN MATHEMATICS

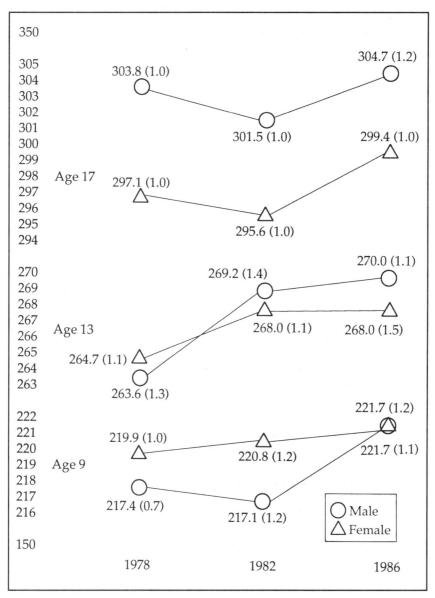

Fig. 13.1. National trends in male and female mathematics achievement

males had significantly higher proficiency levels in all three assessments. That both males and females made significant gains from 1982 to 1986 is encouraging, but the gap between males and females remains a concern.

A closer examination of the proficiency data for the 17-year-olds reveals more about the gender difference. Table 13.2 shows the percentages of 17-

Table 13.2
Trends in Percentage of Females and Males at Age 17 with Mathematics Proficiency at or above a Given Level

Proficiency Level		Assessment		
		1978	1982	1986
350	Males	9.5 (0.5)*	6.7 (0.6)*	8.2 (0.7)*
	Females	5.5 (0.4)*	4.1 (0.4)*	4.5 (0.6)*
300	Males	54.9 (1.2)*	51.9 (1.4)*	54.2 (1.6)*
	Females	48.0 (1.2)*	44.9 (1.3)*	48.1 (1.5)*
250	Males	93.0 (0.5)*	93.9 (0.6)*	96.5 (0.6)*
	Females	91.2 (0.6)*	92.0 (0.5)*	95.5 (0.4)*
200	Males	99.9 (0.0)	99.9 (0.0)	99.9 (0.1)
	Females	99.7 (0.1)	99.9 (0.0)	99.9 (0.1)
150	Males	100.0 (0.0)	100.0 (0.0)	100.0 (0.0)
	Females	100.0 (0.0)	100.0 (0.0)	100.0 (0.0)

Jackknifed standard errors are presented in parentheses.
*Percentages for females and males are significantly different, $p < .05$.

year-old males and females at or above the various proficiency levels for each of the last three assessments. At the lower levels, the percentages of females and males do not differ. However, the percentages of males who attained the higher proficiency levels of 250, 300, and 350 are significantly greater than the percentages of females. This finding suggests that the gender differences in mathematics achievement result from the best males performing at higher levels than the best females.

Content Subscales

Similar to previous assessments, results were computed for five mathematics content-area subscales: Knowledge and Skills, Higher-Level Applications, Measurement, Geometry, and Algebra. Information about the relative achievement of females and males on these subscales gives us a fuller understanding of the nature of the gender differences found in the assessment. Table 13.3 gives the average mathematics proficiency by gender for each of the five subscales.

The results on the Knowledge and Skills and Higher-Level Applications subscales are consistent with those reported for earlier assessments. Both subscales comprise items on whole numbers, common fractions, decimals, and percents. On the Knowledge and Skills subscale, which measures performance of straightforward, routine manipulations, females showed a consistent advantage. This contrasts with the males' advantage on the Higher-Level Applications subscale, which measures the ability to go beyond knowledge, skill, and understanding to identify and implement an appropriate strategy to solve a problem.

GENDER DIFFERENCES IN MATHEMATICS 153

Table 13.3
Average Mathematics Proficiency in Content Area Subscales by Gender at Grades 3, 7, and 11: 1986

	Numbers and Operations: Knowledge and Skills		
	Grade 3	Grade 7	Grade 11
Males	212.8 (1.2)	274.0 (0.8)*	302.7 (0.9)
Females	215.0 (0.9)	279.6 (0.9)*	304.4 (0.7)
	Numbers and Operations: Higher-Level Applications		
	Grade 3	Grade 7	Grade 11
Males	211.8 (0.9)	259.7 (0.6)	308.4 (1.1)*
Females	210.6 (1.0)	258.4 (0.8)	300.4 (0.9)*
	Measurement		
	Grade 3	Grade 7	Grade 11
Males	213.1 (0.9)*	265.8 (0.8)	309.3 (1.1)*
Females	208.8 (0.7)*	264.6 (0.8)	298.5 (1.0)*
	Geometry		
	Grade 3	Grade 7	Grade 11
Males	—	266.7 (0.5)	307.1 (1.0)*
Females	—	265.8 (0.6)	301.1 (0.8)*
	Algebra		
	Grade 3	Grade 7	Grade 11
Males	—	—	303.6 (1.3)
Females	—	—	304.0 (0.9)

*Percentages for males and females are significantly different, $p < .05$.

On the Measurement subscale, males performed better than females at all three grade levels. The differences were significant at grade 3 and grade 11. Table 13.4 contains the results for males and females at all three grade

Table 13.4
Percentage Correct by Gender on Three Measurement Items

Item A	When asked to identify the largest of four metric length units, the following percentages of males and females responded correctly at the three grade levels: *Grade 3:* males: 32; females: 21 *Grade 7:* males: 66; females: 58 *Grade 11:* males: 80; females: 64
Item B1	When asked to select the number of meters that best estimated the height of a common object in their classroom, the following percentages of males and females responded correctly at grades 7 and 11: *Grade 7:* males: 41; females: 27 *Grade 11:* males: 52; females: 36
Item B2	An item parallel to item B1 was given using feet as the unit, with the following results: *Grade 7:* males: 78; females: 67 *Grade 11:* males: 90; females: 76

levels on three measurement items. The first item required students to identify the largest of four metric units of length. The second pair of items required students to estimate the height of a common object in their classroom. The first item of the pair contained metric units and the second contained standard units.

Since none of these items are at a high cognitive level, one wonders why such large gender differences were found. The first two items might suggest that males are more familiar with the metric system, but the differences persisted on the third item, involving standard units. Also, it would seem that males and females would have equal exposure to metric units in school. One possible explanation is that males have more experience with measurement instruments, as shown by results from the NAEP science assessment.

No gender differences were found on the Geometry subscale at grade 7, but at grade 11, males performed significantly better than females. Table 13.5 contains the results for two geometry items that measured understanding of properties of triangles.

Table 13.5
Percentage Correct by Gender on Two Geometry Items Having No Visual Component

Item A	Which of the following sets of numbers CANNOT be lengths of the sides of a triangle?
	4,7,9
	11,9,5 Grade 7: males: 12; females: 7
	8,3,7 Grade 11: males: 28; females: 19
	9,5,3*
	3,5,6
Item B	Which of the following sets CANNOT be measures in degrees of the interior angles of a triangle?
	70,60,50
	100,40,40 Grade 7: males: 18; females: 19
	90,70,20 Grade 11: males: 55; females: 47
	100,60,30*
	140,20,20

*Indicates the correct response.

Given the nonvisual nature of these items, no obvious reason explains these gender differences. However, it was not surprising that some items with a visual component produced differences. Table 13.6 contains two such examples.

The Algebra subscale results do not indicate any gender differences.

GENDER DIFFERENCES IN ATTITUDES

The 1986 assessment contained numerous items assessing attitudes. Attitudes about mathematics and about oneself as a learner of mathemat-

Table 13.6
Percentage Correct by Gender on Two Geometry Items Having a Visual Component

Item A

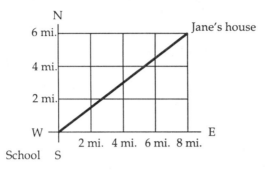

Jane rode 8 miles east, then 6 miles north from her home to school. If she could have gone straight from home to school, how far would she have ridden?

6
8
10*   Grade 11: males: 43; females: 28
14

Item B

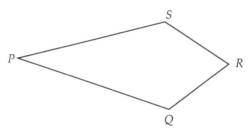

A tiny figure of a polygon like the one above is placed under a microscope. Under the microscope, the polygon looks larger. If the *actual figure* and its *image as seen through the microscope* are compared, what property changes?

The measure of angle Q
The length of segment PS*
The intersection points of lines QR and RS   Grade 11: males: 47; females: 31
The sum of the interior angles

*Indicates the correct response.

ics are important not only because they are outcomes of learning mathematics but also because they potentially influence the learning of new mathematics. Attitudinal data give us clues concerning the origin and development of the achievement and participation differences that were found for males and females. The results will be reported for each dimension of the attitudes measured.

Confidence

Several items measured dimensions of confidence in mathematics, which is the student's perception of his or her ability to perform well on mathematical tasks and to learn new mathematics. Past research on this variable has shown that males generally express higher confidence than do females, even when no achievement differences are present (Fennema and Sherman 1977; 1978). Results from the 1986 NAEP are consistent with this research. Students were asked to express their degree of agreement with the statement "I am good at mathematics." As table 13.7 shows, no sex differences in agreement with this statement were found in grade 3 but in grade 7, significantly more males expressed agreement than females, and this difference was repeated in grade 11.

Table 13.7
Confidence Results for Males and Females

Statement	Percent Strongly Agree or Agree	
I am good at mathematics	Grade 7	Grade 11
Males	63.5 (2.1)*	58.3 (2.5)*
Females	56.8 (2.3)*	48.3 (1.9)*
	Percent Yes	
Are you as good in math as others your age?	Grade 3	Grade 7	Grade 11
Males	62.2 (1.8)	72.9 (1.4)*	71.7 (1.7)*
Females	60.7 (1.6)	66.5 (1.8)*	60.2 (1.8)*

Jackknifed standard errors are in parentheses.
*Male and female percentages are significantly different, $p < .05$.

Another item that measured confidence asked, "Are you as good in math as others your age?" As seen in table 13.7, there were no differences in the percentages of males and females answering "yes" to this question in grade 3, but in grades 7 and 11, significantly more males than females said "yes."

Usefulness

A perception of the usefulness of mathematics is another variable that has been researched because of its possible influence on mathematics participation for males and females (Armstrong and Price 1982; Lantz and Smith 1981). When examined for gender differences, results from selected items that measured students' perceptions of the usefulness of mathematics seem to be contradictory. As table 13.8 shows, in grades 7 and 11, females and males agreed equally with the statement "Most of mathematics has practical use." However, significantly more males than females in grades 7 and 11 expressed agreement with the statement "When you think about

Table 13.8
Usefulness Results for Males and Females

Statement	Percent Strongly Agree or Agree	
Most of mathematics has practical use.	Grade 7	Grade 11
Males	78.6 (1.7)	82.4 (2.9)
Females	80.8 (1.8)	79.5 (2.3)
	Percent Yes	
When you think about what you will do when you are older, do you expect that you will work in an area that requires mathematics?	Grade 7	Grade 11
Males	47.5 (1.3)*	52.8 (1.5)*
Females	40.2 (1.3)*	39.9 (1.7)*

Jackknifed standard errors are in parentheses.
*Male and female percentages are significantly different, $p < .05$.

what you will do when you are older, do you expect that you will work in an area that requires mathematics?" This apparent discrepancy is perhaps explained by the fact that the first question is general in nature and a student can agree with it without expressing any personal commitment to the further study of more mathematics. The second question, however, is more specific and requires a personal response. It would seem that although females recognize the potential usefulness of mathematics, this utility is perceived in the abstract and is not necessarily connected to their goals and ambitions.

Sex-Role Stereotyping

Another potential influence on participation and achievement for males and females is that of sex-role stereotyping (Armstrong and Price 1982). Table 13.9 shows the stereotyping trends for 13- and 17-year-olds for the

Table 13.9
Sex-Role Stereotyping Trends for 13- and 17-Year-Old Males and Females

Statement			Percent Strongly Agree or Agree		
			1978	1982	1986
Mathematics is more for boys than girls.	Males	13	3.4 (0.6)*	4.3 (0.6)*	6.2 (1.0)*
		17	3.0 (0.7)*	3.3 (0.6)*	4.2 (1.0)*
	Females	13	1.7 (0.3)*	2.3 (0.7)*	6.1 (0.9)[a]
		17	1.5 (0.4)*	1.5 (0.4)*	2.0 (0.6)*

Jackknifed standard errors are in parentheses.
*Male and female percentages are significantly different, $p < .05$.
[a] From 1982 to 1986, percentages for females are significantly different, $p < .05$.

past three assessments. In each year, males at both ages have been significantly more likely than females to agree that mathematics is "more for boys."

This result replicates data recently obtained in the United States results from the Second International Mathematics Study (IEA 1984). In that study, students in grades 8 and 12 generally held a positive view of females and mathematics, but when the responses were separated by gender, males were much more likely to stereotype mathematics as a male domain.

It is important to note that in the NAEP data, although the percentages of males and females who stereotyped mathematics were low, significantly more 13-year-old females agreed with the statement in the 1986 assessment than did in 1982. When coupled with the decline in the percentage of 13-year-old females who would like to take more mathematics (48% in 1982 vs. 40% in 1986), this finding might indicate that the impact of intervention efforts to change stereotypic perceptions has begun to erode.

CONCLUSIONS

As was found in the previous two assessments, few gender differences were found in the mathematics achievement of 9- or 13-year-olds. Since 1982 the 9-year-old males made significant gains in proficiency compared to only slight gains by the 9-year-old females. The proficiency of 13-year-old males and females and their relative positions showed little change from 1982 to 1986. Both male and female 17-year-olds showed significant gains in the 1986 assessment, which possibly reflects increased mathematics course taking since the last assessment. However, the gender difference in proficiency for 17-year-olds, especially at the highest proficiency levels, remained basically unchanged. Males are still performing a higher proficiency levels despite only small differences in mathematics course taking. The content areas that contributed most to differences for 17-year-olds were higher-level applications of numbers and operations, measurement, and geometry. Differential experiences with measurement instruments might contribute to the differences found in measurement. Likewise, differences in spatial ability for males and females partially account for the differences found in geometry. However, in both cases a fuller understanding is needed.

Increased awareness, understanding, and intervention efforts over the past decade seem to have done little to change the pattern of gender differences in achievement. Changes in course-taking patterns for both males and females are evident and encouraging, but they are clearly not enough to ensure equal outcomes in either achievement or in attitudes. Females continue to express less confidence in their mathematical ability and a lower perception of the usefulness of mathematics to them in the future. The greater stereotyping of mathematics by males remains troublesome and could be a deterrent for some females questioning their future involvement in mathematics. Equity for males and females in mathematics has not been realized, and efforts to achieve this goal should not be relaxed until all differences in outcomes have disappeared.

The contribution of the NAEP mathematics assessment in documenting gender differences is important, but unfortunately it does little to identify the origins of the differences or the means to eliminate them. Research on the causal connections between such variables as attitudes and achievement-related behaviors will move us in the direction of equal outcomes in mathematics for all females and males.

REFERENCES

Armstrong, Jane M., and Richard A. Price. "Correlates and Predictors of Women's Mathematics Participation." *Journal for Research in Mathematics Education* 13 (March 1982): 99-10.

Fennema, Elizabeth, and Julia A. Sherman. "Sex-related Differences in Mathematics Achievement, Spatial Visualization and Affective Factors." *American Educational Research Journal* 14 (1977): 51-71.

Fennema, Elizabeth, and Julia A. Sherman. "Sex-related Differences in Mathematics Achievement and Related Factors: A Further Study. *Journal for Research in Mathematics Education* 9 (May 1978): 189-203.

International Association for the Evaluation of Educational Achievement. *Second International Mathematics Study Summary Report for the United States.* Champaign, Ill.: U.S. National Coordinating Center, 1984.

Lantz, A. E., & G. P. Smith. "Factors Influencing the Choice of Nonrequired Mathematics Courses." *Journal of Education Psychology* 73 (1981): 825-837.

14

SUMMARY AND CONCLUSIONS

Thomas P. Carpenter   Mary M. Lindquist

THE previous chapters have included discussions of results for specific topics or selected subgroups. In this chapter, we summarize some of the major trends that cut across the content areas and subgroups. Although the fourth assessment targeted grades 3, 7, and 11; trend data were collected from students who were 9, 13, or 17 years old for consistency with previous assessments. Figure 14.1 shows the trends over the past three assessments.

HIGHLIGHTS

• The overall performance of 9-year-olds, which had showed little change from 1973 to 1982, improved significantly between 1982 and 1986.

• The performance of 13-year-olds, which had increased in the late 1970s and early 1980s, leveled off—there was virtually no change from 1982 to 1986.

• For 17-year-olds, the downward trend that characterized their performance in the 1970s was reversed: this group made significant gains between 1982 and 1986.

• The gains made by 17-year-olds paralleled the gains made by 13-year-olds between 1978 and 1982. Since these trends represent the performance of the same age cohorts, one can speculate that the gains registered by 17-year-olds are a continuation of changes that were observed for 13-year-olds four years earlier.

• The trends for 17-year-olds are consistent with the results of other large-scale national testing programs, such as the mathematics portion of the College Board's Scholastic Aptitude Test (SAT). SAT scores declined between 1973 and 1978, leveled off between 1978 and 1982, and increased modestly between 1982 and 1986.

SUMMARY AND CONCLUSIONS

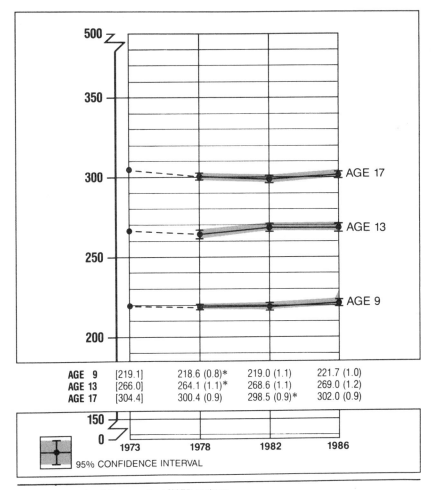

Fig. 14.1. National trends in mathematics achievement for 9-, 13-, and 17-year-olds, 1973–86. (Reprinted from *The Mathematics Report Card: Are We Measuring Up?* Princeton, N.J.: Educational Testing Service, 1988; p. 19.)

COURSE TAKING

Recently a number of states have increased the number of mathematics courses students are required to take for graduation. It is still too early to observe the full impact of this trend on the course-taking patterns of high school juniors, but from 1982 to 1986 there was a slight increase in the number of advanced-level mathematics courses that 17-year-olds reported having taken (see table 14.1). Fewer students are completing their high school mathematics education without taking algebra or geometry, and at all levels of the curriculum enrollment has increased slightly. Despite these indicators of increased mathematics course taking, almost 40 percent of the 17-year-olds reported not having taken a mathematics course beyond Algebra 1.

Table 14.1
Trends in Course Taking

	Algebra 1			Geometry			Algebra 2			Precalculus or Calculus		
	1978	1982	1986	1978	1982	1986	1978	1982	1986	1978	1982	1986
Nation	17	16	18	16	14	17	37	39	40	6	5	7
Male	15	16	17	15	13	15	38	39	39	7	6	8
Female	18	17	18	18	15	18	37	39	40	4	5	5
White	17	15	17	17	15	17	39	41	42	6	5	7
Black	19	20	18	11	10	16	28	29	31	4	4	3
Hispanic	19	21	24	12	12	16	23	24	28	3	3	6

CURRENT ACHIEVEMENT

The most critical question is not whether students' performance has changed over time but whether they are learning what they should be learning. The results of the fourth NAEP assessment indicate that serious gaps exist in students' knowledge and that they are learning a number of concepts and skills only at a superficial level.

About a third of the 7th-grade students and a fourth of the 11th-grade students demonstrated extremely limited knowledge of some basic mathematical concepts and skills. Although they could perform simple whole number calculations, they gave little evidence of knowing the most fundamental concepts of fractions, decimals, or percents. Similarly, they could identify simple geometric figures, make simple measurements, and read simple graphs, but they could not use basic properties of geometric figures, compute areas or volumes, or draw conclusions from graphs and tables.

Learning Concepts and Skills

One of the central issues underlying recent reforms of the mathematics curriculum has been the relative emphasis that should be placed on developing understanding of basic concepts and the teaching of mathematical skills. The modern mathematics movement of the 1960s emphasized understanding, whereas the back-to-basics movement of the 1970s focused on teaching skills. The question, however, is not one of choosing between understanding and skills. Mounting evidence shows that skills cannot be effectively learned in isolation and that students must understand the skills they are learning to apply these skills flexibly in a variety of contexts.

This assessment suggests that many students are not developing an understanding of the concepts underlying the skills they are attempting to learn. For example, the difficulty that 3rd-grade students encounter in adding three-digit whole numbers can be traced to their lack of understanding of place value for three-digit numbers; older students' difficulties with fractions, decimals, and percents reflect serious gaps in their knowledge of the underlying concepts.

Performance on the items in table 14.2 illustrates the limits of students' knowledge of the basic meanings of fractions and decimals. Many students who are successful at routine, frequently encountered calculations have difficulty on questions that do not involve standard calculations presented in a familiar context, even when they involve basic number concepts.

Table 14.2.
Basic Meanings of Fractions and Decimals

	Percent Correct	
Item	Grade 7	Grade 11
A. $5\frac{1}{4}$ is the same as	47	44
$\quad 5 + \frac{1}{4}$ *		
$\quad 5 - \frac{1}{4}$		
$\quad 5 \times \frac{1}{4}$		
$\quad 5 \div \frac{1}{4}$		
B. Write .037 as a fraction:	48	58

*Indicates correct response.

Results on the measurement items in table 14.3 illustrate the errors resulting from students' limited understanding of procedures when they are given a problem in a context that is slightly different from the usual one. The first item has a familiar context, and over 75 percent of the 7th-grade students could use a ruler to measure to the nearest quarter inch. In contrast, on the second item in which the beginning of the segment is not at zero, fewer than half of the 7th-grade students gave the correct answer that the length of this segment was 5 units.

The items in table 14.4 illustrate the difficulty that students have in generalizing the procedures they have learned. Almost half of the 7th-grade students could calculate the area of a rectangle, but only 14 percent could

Table 14.3
Understanding Measuring with a Ruler

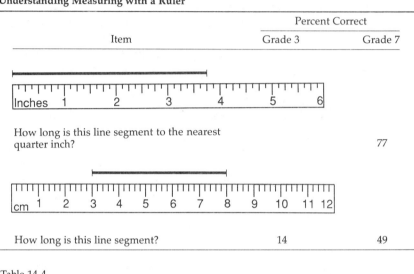

Item	Percent Correct	
	Grade 3	Grade 7
How long is this line segment to the nearest quarter inch?		77
How long is this line segment?	14	49

Table 14.4
Generalizing the Formula for the Area of a Rectangle

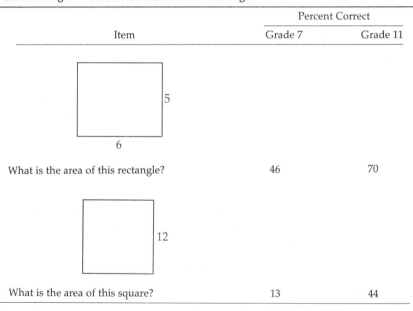

Item	Percent Correct	
	Grade 7	Grade 11
What is the area of this rectangle?	46	70
What is the area of this square?	13	44

apply this knowledge to finding the area of a square. Even though almost all students knew that the sides of a square are equal, they failed to see how they could use this fact and the procedure for finding the area of a rectangle to determine area of the square.

SUMMARY AND CONCLUSIONS

The items in table 14.5 offer another example of the procedural orientation of many students' knowledge. The majority of 11th-grade students who had completed one or two years of algebra could perform the symbolic manipulations involved in solving equations or simplifying expressions. Very few, however, could identify the relationships between variables that were implied by the equation of the third item. They could manipulate the variables, but they did not understand what they represented.

Table 14.5
A Procedural Orientation

	Percent Correct	
Item	Algebra 1	Algebra 2
A. Solve: $6x + 5 = 4x + 7$	83	91
B. Simplify: $9(1 + 5x) + 13$	74	87
C. $x - y > x + y$ implies— $y < 0$ $x > 0$ $x = 0$ $x = y$	38	50

Problem-solving Performance

Recommendations made by the National Council of Supervisors of Mathematics in 1978 and the National Council of Teachers of Mathematics in 1980 called for a renewed emphasis on problem solving. Although it may be too early to find an effect of these recommendations, problem-solving performance, except for one-step routine word problems, remains low. The following results illustrate some of the various types of difficulties students encountered.

The pair of problems in table 14.6 illustrates the difficulty students in the 3rd and 7th grades have when a problem contains extra information. In each

Table 14.6
Problems with and without Extraneous Numbers

	Percent Correct	
Item	Grade 3	Grade 7
A. Jake has 32 toy trucks. He buys 45 more. How many does he have in all?	88	95
B. At the store, a package of screws costs 30¢, a roll of tape costs 35¢, and a box of nails costs 20¢. What is the cost of a roll of tape and a package of screws?	58	78

of these problems, students needed only to add two numbers, but in the second one, they first had to select which two numbers to add. The most common error made was to add all three numbers.

The results of the pair of problems in table 14.7 contrast performance on a one-step problem and a two-step problem involving a percent. The slightly more difficult computation in the second item was not the cause of the difference in the results. Many students found the amount of interest but not the total amount to be repaid.

Table 14.7
One- and Two-Step Problems

Item	Percent Correct	
	Grade 7	Grade 11
A. Robin's baseball team lost 75% of its games last year. If they played 40 games, how many games did they lose?	38	68
B. Jesse borrowed $650 for one year to buy a motorcycle. If she paid 18% simple interest on the loan, what was the total amount she repaid?	9	37

The pair of problems in table 14.8 contrasts 11th-grade students' performance on a problem for which a standard procedure was sufficient and one for which understanding of the concept of average was needed.

Table 14.8
Extending Standard Procedures

Item	Percent Correct
	Grade 11
A. Here are the ages of six children: 13, 10, 8, 5, 3, 3 What is the average age of these children?	72
B. Edith has an average (mean) score of 80 on five tests. What score does she need on the next test to raise her average to 81?	24

The problem in table 14.9 illustrates 7th and 11th-grade students' performance on nonroutine problems. No set procedure exists for solving this problem; that is, one cannot simply add, subtract, multiply, divide, or use combinations of these operations. One must understand the problem and use some reasoning skills to solve it.

The poor performance of the youth of our nation on multistep and nonroutine problems is one of the reasons the National Council of Teachers

SUMMARY AND CONCLUSIONS

Table 14.9
Nonroutine Problem

Item	Percent Correct	
	Grade 7	Grade 11
Suppose you have 8 coins and you have at least one of a quarter, a dime, and a penny. What is the *least* amount of money you could have?	23	43

of Mathematics continues to emphasize the importance of problem solving. In its *Curriculum and Evaluation Standards for School Mathematics,* NCTM's first recommendation for all grades addresses the goal of having students become good problem solvers.

TECHNOLOGY

The rapid growth in technological tools and procedures provides an impetus for change in what mathematics is to be learned and the way in which it is taught. It is interesting to note that computers are accepted and used more than calculators.

HIGHLIGHTS

- Most students have calculators available at home, but relatively few schools provide calculators for use in the classroom and over half of the students never use calculators in school.
- Since 1978, performance on items for which calculators were available has declined significantly for all age groups.
- Students did significantly better on straightforward computational items when calculators were available than when they were not.
- The percentage of 17-year-olds who have completed coursework in computer programming has steadily increased over the last three assessments: approximately one-third now report having had some coursework in computer programming.
- Nearly half of the 13-year-olds and over half of the 17-year-olds have access to computers to learn mathematics, also a sharp increase from 1978. This result must be viewed cautiously because it is not clear how often students have access to computers or for what particular purposes.

MINORITIES AND MATHEMATICS

Over the last decade, black and Hispanic students have made steady gains in achievement in mathematics at all grade levels. Black students at ages 9, 13, and 17 have shown significant improvement across all four mathematics assessments. Hispanic students at age 9 have also shown a slight improvement over the four assessments. At age 13, performance improved significantly from 1978 to 1982, then leveled off between 1982 and 1986. At age 17, after showing little change from 1973 to 1978, performance improved from 1978 to 1986.

The gains of black and Hispanic students on the whole, have been greater and more consistent than the gains of white students, so the gap between the performance of white students and that of black and Hispanic students has been narrowing. More progress is necessary, however, since significant differences remain between the performance of white students and minority students.

Black and Hispanic students also reported significant gains in enrollment in high school mathematics courses between 1982 and 1986. In 1986 the number of black 17-year-olds who reported taking no high school mathematics courses beyond general or business mathematics declined by 4 percent and enrollment in high school mathematics courses showed a comparable gain. Hispanic students showed an even greater change in the pattern of course taking. From 1982 to 1986, the number of Hispanic 17-year-olds who reported taking no mathematics beyond general mathematics declined by over 12 percent and comparable gains were reported for enrollment in advanced mathematics courses. Despite these gains, serious inequities remain in the course enrollment reported by black and Hispanic 17-year-olds. In the eleventh grade, black and Hispanic students reported taking almost a full year less of advanced mathematics than white students.

In the last decade, black and Hispanic students have made important gains in achievement and participation in mathematics courses, but there is much room for improvement. Programs designed to help increase the educational opportunity and mathematics proficiency of minority students must continue to be a high priority for educators and policymakers. The gains suggest that programs implemented over the last ten to twenty years to improve the performance of minority students are having an effect, but to provide real equity in educational opportunity, even greater efforts are needed.

GENDER AND MATHEMATICS

Previous assessments have found few gender-related differences in mathematics achievement at ages 9 and 13, but at age 17 there have been small but significant differences, with males scoring higher than females.

The same pattern was found in the 1986 results.

Little difference was found in the mathematics courses that male and female 17-year-olds reported taking through Algebra 2. About 3 percent more males than females, however, reported enrolling in advanced courses, such as precalculus or calculus. Almost twice as many males as females reported taking a course in computer programming.

Little in the data on achievement or enrollment in high school mathematics classes explains the significant differences between males' and females' participation in mathematical careers; however, as early as grade 7 significantly more males than females responded that they were likely to enter a career that used mathematics; this disparity had increased by grade 11.

CONCLUSIONS

Following the broad declines in student achievement in the 1970s, it appears that mathematics achievement has taken a slight upturn in the 1980s. This trend is not a cause for complacency, however, as student achievement at all age levels shows serious deficiencies.

The emphasis on computational skills that has generally characterized mathematics instruction has left many students with serious gaps in their knowledge of underlying concepts. As a result, students have not learned many advanced skills and frequently they cannot apply the skills they have learned. Moreover, the skills they have learned are in danger of becoming obsolete as technological advances alter the mathematics that adults need to function productively in society. Students' lack of flexibility in applying the mathematics they have learned may leave them unable to adapt to these demands.

The curricular reforms proposed by the National Council of Teachers of Mathematics in the *Curriculum and Evaluation Standards for School Mathematics* call for a reorientation of the school mathematics curriculum so that a greater emphasis is placed on helping students become mathematical problem solvers and communicate and reason mathematically. The results of the fourth NAEP mathematics assessment indicate that these are areas most critically in need of reform.

BIBLIOGRAPHY

Fourth Mathematics Assessment, 1985–86

NAEP Publications

Benton, Albert E. *Implementing the New Design: The NAEP 1983–84 Technical Report.* 15–TR–20. Princeton, N.J.: Educational Testing Service, March 1987.

Dossey, John A., Ina V. S. Mullis, Mary M. Lindquist, and Donald L. Chambers. *The Mathematics Report Card: Are We Measuring Up?* Princeton, N.J.: Educational Testing Service, 1988.

NAEP. *Math Objectives: 1985–86 Assessment.* 17–M–10. Princeton, N.J.: Educational Testing Service, 1986.

NCTM Publications

Brown, Catherine A., Thomas P. Carpenter, Vicky L. Kouba, Mary M. Lindquist, Edward A. Silver, and Jane O. Swafford. "Secondary School Results for the Fourth NAEP Mathematics Assessment: Discrete Mathematics, Data Organization and Interpretation, Measurement, Number, and Operations." *Mathematics Teacher* 81 (April 1988): 241–48.

———. "Secondary School Results for the Fourth NAEP Assessment: Algebra, Geometry, Mathematical Methods, and Attitudes." *Mathematics Teacher* 81 (May 1988): 337–47, 397.

Carpenter, Thomas P., Mary M. Lindquist, Catherine A. Brown, Vicky L. Kouba, Edward A. Silver, and Jane O. Swafford. "Results of the Fourth NAEP Mathematics Assessment: Trends and Conclusions." *Arithmetic Teacher* 35 (December 1988): 38–41.

Kouba, Vicky L., Carpenter A. Brown, Thomas P. Carpenter, Mary M. Lindquist, Edward A. Silver, and Jane O. Swafford. Results of the Fourth NAEP Assessment of Mathematics: Number, Operations, and Word Problems." *Arithmetic Teacher* 35 (April 1988): 14–19.

———. "Results of the Fourth NAEP Assessment of Mathematics: Measurement, Geometry, Data Interpretation, Attitudes, and Other Topics." *Arithmetic Teacher* 35 (May 1988): 10–16.

Silver, Edward A., Mary M. Lindquist, Thomas P. Carpenter, Catherine A. Brown, Vicky L. Kouba, Jane O. Swafford. "The Fourth NAEP Mathematics Assessment: Performance Trends and Results and Trends for Instructional Indicators." *Mathematics Teacher* 81 (December 1988): 720–727.

Other Publications

Lindquist, Mary M., Thomas P. Carpenter, Catherine A.. Brown, Vicky L. Kouba, Edward A. Silver, and Jane O. Swafford. "NAEP: Results of the Fourth Mathematics Assessment." *Education Week* (15 June 1988): 28–29.

Third Mathematics Assessment, 1981–82

NAEP Publications

National Assessment of Educational Progress. *The Third National Mathematics Assessment: Results, Trends and Issues.* 13–MA–01. Denver: Education Commission of the States, 1983.

———. *Mathematics Objectives, 1981–82 Assessment.* Denver: Education Commission of the States, 1981.

NCTM Publications

Carpenter, Thomas P., Mary Montgomery Lindquist, Westina Matthews, and Edward A. Silver. "Results of the Third NAEP Mathematics Assessment: Secondary School." *Mathematics Teacher* 76 (December 1983): 652–59.

Lindquist, Mary Montgomery, Thomas P. Carpenter, Edward A. Silver, and Westina Matthews. "The Third National Mathematics Assessment: Results and Implications for Elementary and Middle Schools." *Arithmetic Teacher* 31 (December 1983): 14–19.

Matthews, Westina, Thomas P. Carpenter, Mary Montgomery Lindquist, and Edward A. Silver. "The Third National Assessment: Minorities and Mathematics." *Journal of Research in Mathematics Education* (15 March 1984): 165–71.

Other Publications

Carpenter, Thomas P., Westina Matthews, Mary Montgomery Lindquist, and Edward A. Silver. "Achievement in Mathematics: Results from the National Assessment." *Elementary School Journal* 84 (May 1984): 485–497.

Second Mathematics Assessment, 1977–78

NAEP Publications

National Assessment of Educational Progress. *Changes in Mathematical Achievement: 1973–78.* 09-MA-01. Denver: Education Commission of the States, 1979.

———. *Mathematical Applications.* 09-MA-03. Denver: Education Commission of the States, 1979.

———. *Mathematical Knowledge and Skills.* 09-MA-02. Denver: Education Commission of the States, 1979.

———. *Mathematical Objectives: Second Assessment.* Denver: Education Commission of the States, 1978.

———. *Mathematical Understanding.* 09-MA-04. Denver: Education Commission of the States, December 1979.

———. *The Second Assessment of Mathematics, 1977–78: Released Exercise Set.* Denver: Education Commission of the States, 1979.

NCTM Publications

Anick, Constance M., Thomas P. Carpenter, and Carol Smith. "Minorities and Mathematics: Results from National Assessment." *Mathematics Teacher* 74 (October 1981): 560–66.

Bestgen, Barbara J. "Making and Interpreting Graphs and Tables: Results and Implications from National Assessment." *Arithmetic Teacher* 28 (December 1980): 26–29.

Carpenter, Thomas P., Mary Kay Corbitt, Henry S. Kepner, Jr., Mary Montgomery Lindquist, and Robert E. Reys. "Calculators in Testing Situations: Results and Implications from National Assessment." *Arithmetic Teacher* 28 (January 1981): 34–37.

———. "The Current Status of Computer Literacy: NAEP Results for Secondary Students. *Mathematics Teacher* 73 (December 1980): 669–73.

———. "Decimals: Results and Implications from National Assessment." *Arithmetic Teacher* 28 (April 1981): 34–37.

———. "NAEP Note: Problem Solving." *Mathematics Teacher* 73 (September 1980): 427–33.

———. "Results and Implications of the Second NAEP Mathematics Assessment: Elementary School." *Arithmetic Teacher* 27 (April 1980): 10–12, 44–47.

———. "Results of the Second NAEP Mathematics Assessment." *Mathematics Teacher* 73 (May 1980): 329–38.

———. "Solving Verbal Problems: Results and Implications from National Assessment." *Arithmetic Teacher* 28 (September 1980): 8–12.

———. "Students' Affective Response to Mathematics: Results and Implications from National Assessment." *Arithmetic Teacher* 28 (October 1980): 34–37, 52–53.

———. "Students' Affective Responses to Mathematics: Secondary School Results from National Assessment." *Mathematics Teacher* 73 (October 1980): 531–39.

———. "What Are the Chances of Your Students Knowing Probability?" *Mathematics Teacher* 74 (May 1981): 342–44.

———. "Results from the Second Mathematics Assessment of the National Assessment of Educational Progress." Reston, Va.: National Council of Teachers of Mathematics, 1981.

Fennema, Elizabeth, and Thomas P. Carpenter. "Sex Differences in Mathematics: Results from National Assessment." *Mathematics Teacher* 74 (October 1981): 554–59.

Hiebert, James. "Units of Measure: Results and Implications from National Assessment." *Arithmetic Teacher* 28 (February 1981): 38–43.

Hirstein, James. "The Second National Assessment in Mathematics: Area and Volume." *Mathematics Teacher* (December 1981): 704–8.

Kerr, Donald. "A Geometry Lesson from National Assessment." *Mathematics Teacher* 74 (January 1981): 27–32.

McKillip, William D. "Computational Skill in Division: Results and Implications from National Assessment." *Arithmetic Teacher* 28 (March 1981): 34–37.

Post, Thomas R. "Fractions: Results and Implications from National Assessment." *Arithmetic Teacher* 28 (May 1981): 26–31.

Rathmell, Edward C. "Concepts of the Fundamental Operations: Results and Implications from National Assessment." *Arithmetic Teacher* 28 (November 1980): 34–37.

Other Publications

Carpenter, T. P., M. K. Corbitt, H. S. Kepner, Jr., M. M. Lindquist, and R. E. Reys. "National Assessment: Implications for the Curriculum of the 1980s." In *Research in Mathematics Education: Implications for the 80s*, edited by Elizabeth Fennema. Washington, D.C.: Association for Supervision and Curriculum Development, 1981.

———. "A Perspective of Students' Mastery of Basic Skills." In *Selected Issues in Mathematics Education*, edited by Mary Montgomery Lindquist. Chicago: National Society for the Study of Education, 1980.

———. "An Interpretation of the Results of the Second NAEP Mathematics Assessment." In *Education in the 80s: Mathematics*, edited by Shirley Hill. Washington, D.C.: National Education Association, 1982.

———. "Problem Solving in Mathematics: National Assessment Results." *Educational Leadership* 37 (April 1980): 562–63.

First Mathematics Assessment, 1972–73

NAEP Publications

National Assessment of Educational Progress. *Consumer Math: Selected Results from the First National Assessment of Mathematics*. 04-MA-02. Denver: Education Commission of the States, June 1975.

———. *The First National Assessment of Mathematics: An Overview*. 04-MA-00. Denver: Education Commission of the States, October 1975.

———. *Math Fundamentals: Selected Results from the First National Assessment of Mathematics*. 04-MA-01. Denver: Education Commission of the States, January 1975.

———. *Mathematics Objectives*. Ann Arbor, Mich.: NAEP, 1970.

———. *Mathematics Technical Report: Exercise Volume*. 04-MA-20. Denver: Education Commission of the States, February 1977.

———. *Mathematics Technical Report: Summary Volume*. 04-MA-21. Denver: Education Commission of the States, September 1976.

NCTM Publications

Carpenter, Thomas P., Terrence G. Coburn, Robert E. Reys, and James W. Wilson. "Notes from National Assessment: Addition and Multiplication with Fractions." *Arithmetic Teacher* 23 (February 1976): 137–42.

———. "Notes from National Assessment: Basic Concepts of Area and Volume." *Arithmetic Teacher* 22 (October 1975): 501–7.

———. "Notes from National Assessment: Estimation." *Arithmetic Teacher* 23 (April 1976): 296–302.

———. "Notes from National Assessment: Perimeter and Area." *Arithmetic Teacher* 22 (November 1975): 586–90.

———. "Notes from National Assessment: Processes Used on Computational Exercises." *Arithmetic Teacher* 23 (March 1976): 217–22.

———. "Notes from National Assessment: Recognizing and Naming Solids." *Arithmetic Teacher* 23 (January 1976): 62–66.

———. "Notes from National Assessment: Word Problems." *Arithmetic Teacher* 23 (May 1976): 389–93.

———. "Research Implications and Questions from the Year 04 NAEP Mathematics Assessment." *Journal for Research in Mathematics Education* 7 (November 1976): 327–36.

———. "Results and Implications of the NAEP Mathematics Assessment: Elementary School." *Arithmetic Teacher* 22 (October 1975): 438–50.

———. "Results and Implications of the NAEP Mathematics Assessment: Secondary School." *Mathematics Teacher* 68 (October 1975): 453–70.

———. *Results from the First Mathematics Assessment of the National Assessment of Educational Progress.* Reston, Va.: National Council of Teachers of Mathematics, 1978.

———. "Subtraction: What Do Students Know?" *Arithmetic Teacher* 22 (December 1975): 653–57.

Martin, Wayne L., and James W. Wilson. "The Status of National Assessment in Mathematics." *Arithmetic Teacher* 21 (January 1974): 49–53.

Other Publications

Reys, Robert E. "Consumer Math: Just How Knowledgeable Are U.S. Young Adults?" *Phi Delta Kappan* (November 1976): 258–60.